WHY SCIENCE?

To know, to Understand, and to Rely on Results

WHY SCIENCE?

To know, to Understand, and to Rely on Results

Roger G. Newton

Indiana University, USA

 **World Scientific**

NEW JERSEY · LONDON · SINGAPORE · BEIJING · SHANGHAI · HONG KONG · TAIPEI · CHENNAI

Published by

World Scientific Publishing Co. Pte. Ltd.

5 Toh Tuck Link, Singapore 596224

USA office: 27 Warren Street, Suite 401-402, Hackensack, NJ 07601

UK office: 57 Shelton Street, Covent Garden, London WC2H 9HE

British Library Cataloguing-in-Publication Data
A catalogue record for this book is available from the British Library.

WHY SCIENCE?
To Know, to Understand, and to Rely on Results

ISBN-13 978-981-4397-33-9 (pbk)
ISBN-10 981-4397-33-4 (pbk)

Typeset by Stallion Press
Email: enquiries@stallionpress.com

Printed in Singapore by World Scientific Printers.

Contents

Introduction   **vii**

1. We Want to Know   **1**

    Looking at the Heavens   4
    Exploratory Voyages   8

2. We Want to Understand   **25**

    Charles Darwin   36
    Gregor Mendel   43
    Louis Pasteur   51
    Michael Faraday   59
    Max Planck   65
    Enrico Fermi   68

3. Science   **71**

    Chemistry as the Fundamental Science   71
    How Physics Became Most Fundamental   74
    On Reductionism   81

References and Further Reading   **85**

Illustration Credits   **87**

Index   **89**

Introduction

The purpose of this book is to trace the development of humanity's desire to know and understand the world around us through the various stages of its history up to the present. Today science is almost universally recognized — at least in the Western world — as the most reliable way of knowing.

Beginning with the awe-inspiring spectacle of what we see with regularity in the sky, cultures that esteemed knowledge obtained by observation rather than by relying on the authority of tradition left us with a legacy that valued understanding rather than mere knowing. I relate how this would eventually evolve into what we now call science, though the word is of relatively recent usage. Fleshing out this history, I describe in the first chapter some of the large-scale exploration of the Earth's surface, long voyages undertaken mostly by Vikings, Chinese, Spaniards, and Portuguese, into the then-unknown interiors of the American and African continents, to the North poles and South poles and into the depths of the oceans, primarily for the sake of knowledge. After the invention of the telescope, I show how visual exploration turned to the surfaces of the Moon and Mars. Eventually, rockets even took explorers to the Moon.

In Chapter 2 we turn to the scientific way of knowing and why it is valued. I take six important scientists, Charles Darwin, Gregor Mendel, Louis Pasteur, Michael Faraday, Max Planck, and Enrico Fermi, and their contributions to serve as exemplars to make the argument concrete.

In the third chapter I finally describe the development that has made physics the most fundamental discipline of the sciences. At the beginning of the nineteenth century, the great chemist Humphry Davy had argued, plausibly, in public lectures that chemistry was the basic science, the foundation of all the life sciences. However, when later during that century and the next, physics began to take the existence of atoms seriously and (with the help of quantum mechanics) even began to explore and successfully describe the structure of their interiors, chemistry turned out to depend heavily on physics. What is more, nuclear physics, particle physics, and the theory of relativity served to extend the reach of physics to all the celestial objects and their motions, making physics the foundation of cosmology as well. It has become clear, I argue, that of all the disciplines of science, physics is both the most basic and the most wide–reaching. Finally, I discuss the controversial question of whether this conclusion commits the sin of reductionism, and if so, why there is nothing wrong with that.

I am much indebted to my wife Ruth for valuable editorial assistance.

1

We Want to Know

Curiosity is the essence of life. Animals cannot live without knowing what is of immediate use to them and what is needed for their survival: where to find food, how to avoid predators, where to find mates, etc. However, the human species differs from other animals because we thirst for knowledge that reaches far beyond our personal needs. We look around us and we wonder. We wonder aimlessly and passively about our surroundings and about what we observe both near and far, but we want to understand it all. Indeed, we fear the unknown. This sense of wonder and the urgent desire for understanding not only makes us human, but is also one of the foundation stones of civilization. The satisfaction of our curiosity becomes easier and more proficient when the search for answers and their retention in a generally accessible memory is pursued in an organized fashion, helping to transform inchoate groups into the grand historical confederations we call civilizations, such as the Chinese, the Babylonian, the Egyptian, the Mayan, the Indian, the ancient Greek, the contemporary Western civilizations, and perhaps others that have left no record.

While the various known historical civilizations had much in common, they also differed fundamentally in many respects. One of the basic differences among some of them was that in their pursuit of knowledge they tended to focus their attention on different aspects of their environment; what is more, they searched for such knowledge in different ways. The aspects of their surroundings and the perceived features of the world which their thinkers and philosophers considered important, worth studying deeply and worth teaching to their disciples to be remembered, differed from one civilization to another as well as sometimes from one epoch to another. Their ideas about how to go about such studies often differed profoundly as well.

The Confucian philosophy of the sixth century BCE in China, whose long-term influence on Chinese thinking is hard to over-estimate, focused primarily on the conventional moral and social organization of traditional Chinese feudal society — its ranks and privileges — and how to perpetuate it. Confucian philosophers paid no attention at all to nature and what might be found there by observation. China had to wait three hundred years until the advent of the Taoist mystics, opposed to Confucian philosophers, who realized that knowledge obtained by paying close attention to the external world was worth acquiring and thinking about in a systematic way. "Taoism was the only system of mysticism which the world has ever seen which was not profoundly anti-scientific" the historian of Chinese civilization, Joseph Needham, cites as a 'well-known comment.'

Religious beliefs also often strongly influenced decisions about what was worth studying and knowing. In the Indian civilization it was Prince Gautama, the Buddha, whose rejection of all superstition and whose respect for reason and rational truth became enormously influential as the foundation of a religion. However, since the Buddha had neither interest in nor curiosity about nature, his thinking retarded attention to the external world in the Indian civilization. Only Jainism, a relatively minor offshoot of Buddhism that was mostly oriented toward ethical and behavioral issues, paid close attention to nature as an incidental sideline.

In Christian Europe during the Middle Ages, when the teaching of educated classes was primarily in the hands of the Church, the acquisition of new scientific knowledge was very slow. Nevertheless what there was of such knowledge, which was mostly inherited from the ancient Greeks, tended to be well preserved and encouraged, notwithstanding such scandalous incidents as the trials of Giordano Bruno and Galileo Galilei. The Islamic world, which initially experienced a great cultural flowering of new contributions to science and mathematics, achieving important advances that far surpassed contemporary Christian Europe's accomplishments, later found itself in a great decline in these areas of knowledge. The reasons for this drastic change remain largely unknown. On the other hand, the Copernican revolution in astronomy, followed by Kepler's much more accurate description of the motion of the planets, was regarded as philosophically revolutionary and initially strongly resisted only in Christian Europe. Neither the Muslims, the Chinese, nor the Hindus were philosophically or religiously offended by the new ideas but regarded them merely as an improved system for constructing a more accurate calendar.

In addition to the difference between the subjects of the knowledge sought, there is also an essential difference between two ways of acquiring and pursuing knowledge: on one hand there is what was later called 'book learning,' the knowledge based on the authority of the written or orally transmitted word, and on the other hand there is knowledge based on independent observations of the world. The contrast between these two manners of seeking knowledge also characterized the conflict between Confucianism and Taoism in China. The former tended to rely mostly on rules written down or orally transmitted over the years, while the latter directed its attention outward, to the world of nature as we see it.

All cultures, of course, had their ancient myths that were regarded as historical knowledge, the veracity of which was taken for granted. In Mesopotamia it was the story of Gilgamesh transmitted and memorized as a long poem; for the ancient Greeks it was Homer with the *Iliad*, the *Odyssey*, and the purported history of the Trojan War;

the Hindus had the *Bhagavad Gita*; the much later legends of the Teutonic tribes were contained in the myths of the *Nibelungen*, those of Beowulf and the fables of dragons; and the Scandinavian people had their tales about the Vikings and their exploits. The three Abrahamic religions, Judaism, Christianity, and Islam, are based on such stories as the creation of the world and the great flood, which are still regarded as known facts by many people.

However, it is clear that only knowledge based on observation rather than authority, myth, or ancient tales eventually led to what we now call science and all its consequences. Conflicts between the two kinds of knowledge, one based on traditional authority, such as the Bible or the Quran, the other empirically acquired and verified, are causing strong social enmity to this day, as continuing fights over Darwinism attest. So let us focus on knowledge gained by observing the world around us.

Looking at the Heavens

The aspects of nature that are most likely to attract the attention of any close observer are surely those phenomena that are regularly repeated. The most striking of these we see in the heavens. The Sun rises every morning, slowly makes its way across the sky along a similar path every day, and sets in the evening. At night, we see the stars and the Moon with its changing phases moving similarly in the sky with fascinating regularity.

So far as is known, the Babylonians seem to have been the first to go beyond sheer wonder, making detailed observations of the heavenly phenomena and recording them. The first instrument employed for this purpose was the gnomon, a simple stick fixed vertically in the earth, or a large stone column with a clear open area around it. The point was to observe the motion of the shadow cast by the slim object on the ground. As the Sun rose and moved across the sky, the length of the shadow, long and pointing toward the west in the morning, shortened to a minimum in the middle of the day, before lengthening and pointing toward the east in the evening, providing a useful compass more accurate than simply looking up at the sun. Not only that, the shadow turned and pointed in a rotating direction in the course of the day — Babylon was not located on the equator, where the shadow would change in length but not rotate — creating a way to measure the time of day for the first time. Even the rudimentary knowledge provided by such a simple instrument as the gnomon served to give useful information that went beyond the vague notions of the progress of time that of course must have existed long before. More convenient ways of using the motion of a shadow for measuring time were soon found and the first sun clocks constructed, followed by water clocks that would function even at night since they measured time by an even outflow of water.

Closer observation, however, showed that the shadow cast by a gnomon was not quite the same from one day to another. Not only did its directions in the morning and evening vary somewhat depending on the season, but so did its minimal length in the middle of the day. In other words, the shadow of the gnomon could not only be used as a measure of time during the day, but also to tell the time of year, how far along the spring was, how much time was left until crops had to be planted, how close the harvesting season was, and other such pieces of very useful and practical information.

The early pursuit and acquisition of astronomical knowledge was thus motivated in large part by its usefulness for the measurement of time and the construction of a calendar. The phases of the Moon in particular lent themselves to the recording of a monthly periodicity, the repetition of which would lead to a measure of the time of year and a calendar based on it, to predict seeding and harvesting seasons, and to keep records of important events. However, the discovery that a year defined by twelve lunar months did not quite correspond to the periodicity of the seasons made it necessary to not only be quite accurate with astronomical measurements but also to introduce complicated corrections so as to maintain the usefulness of the calendar. The Babylonians became extremely skilled at astronomical observations and their recording. The later Greek civilization, which was much less proficient at accurate observations, relied on the old Babylonian records for centuries. (However, a detailed record of the positions of all astronomical objects at a specified date would later be called by the Greek name "ephemeris".) In China too, astronomy was pursued and in fact mandated by imperial order for the specific purpose of constructing and maintaining a reliable calendar.

Observation of the night sky with its many stars, some of them much more prominently visible than others lent itself to practical purposes too. A few, when carefully observed, noticeably moved from one night to the next, while others were fixed. Ships at sea could determine directions by locating the North Star, while Egyptian agriculture was assisted by observing the time of the rising of the star Sirius, the dog star, which served to predict the eagerly awaited fructifying flooding of the Nile. It should not, however, be concluded that astronomy served exclusively utilitarian purposes.

Gazing at the night sky with its many stars of various magnitudes inspired the construction of imaginary constellations of groups resembling real or mythical animals, such as Ursa Major, the Great Bear (also called the Big Dipper), Lupus, the Wolf, and many others such as Gemini, the Twins. What is more, awed by what they saw in the heavens and letting their imagination wander, people made up fanciful tales of the influence of the heavens upon human fates. For millennia the study of astronomy remained inextricably intertwined with its use for astrology. Even the early practice of medicine was often linked to astrological beliefs.

Not that all of astronomy's mythological use was as fanciful as that. Serious study of long-range periodicities that were discovered in the course of time allowed astronomers to predict the occurrence of such awesome, even frightening, phenomena as eclipses of the sun and the moon, which sometimes had fateful, down-to-earth consequences. The oracle of Delphi heaped extravagant praise on the Ionian philosopher Thales, elevating him to be one of the seven legendary Greek Wise Men on the basis of his no doubt apocryphal successful prediction of the solar eclipse of 585 BCE. It happened that this eclipse occurred in the middle of a battle between the armies of the Persians and the Lydians and the sudden darkness so overawed the two kings that they regarded it as a signal from the gods and made peace. Heavenly occurrences thus did sometimes have a direct influence on important earthly events. Of course the claims of astrology were much more far-reaching than that, pretending guidance of individual lives and the happening of events by movements of the stars. Belief in such imaginative claims of knowledge of influences for which there was really no evidence would last for a very long time, even to this day.

The same Ionian philosopher, Thales, acclaimed by legend, in fact deserves enormous amount of credit as the first known man who might be called a proto-scientist. What distinguished him from the earlier Babylonian and Egyptian astronomers, who assembled vast lists of accurate observations but made no attempts to understand or explain them, was the fact that he always wanted to understand whatever was seen, and he forever searched for the fundamental principles underlying whatever regularity was observed. Even when he solved practical problems by means of geometry, such as estimating the distance of a ship from the shore or the height of a building, he was not satisfied by finding a useful geometrical trick but insisted on searching for the general rule on which it was based.

Thales's influence upon Greek thought was all-pervasive. The fact that a triangle formed by a string with sides whose lengths were three, four, and five units long included a right angle between the two shorter sides is believed to have been long known to the Egyptians and used for land surveillance, but it took Pythagoras to prove his eponymous well-known theorem as a general truth. Greek science owes Thales a very large debt.

The two greatest Greek philosophers who studied the origin and purpose of knowledge were of course Plato and Aristotle, and from the fourth century BCE on they exerted a very long-lasting influence on Western thought. 'Let no one enter here who is ignorant of mathematics' proclaimed a sign at the entrance to Plato's Academy, and while his philosophy had a largely negative effect on later science, many modern mathematicians are guided by it even still. His teaching discounted all knowledge obtained by the senses and valued only insights arrived at by pure thought. Information derived from observation of the world provided us with a mere shadow of the 'real'; the underlying 'ideal form' could be grasped by thinking alone. Only mathematics deals with the truth residing in a universe accessible to pure thought and Plato prized it beyond all other forms of knowledge. It took more than two thousand years for science to shed this unproductive idea.

Fundamentally disagreeing with his revered teacher, Plato's student Aristotle was convinced that 'by nature all men desire to know. We all take delight in what our senses present to us.' Contrary to Plato, he taught his students that only observation could lead to reliable information about the world. His beneficial imprint on Western thought would last for millennia until, in place of a clear understanding of his underlying philosophy, it degenerated into sterile book learning of the conclusions he had reached on the basis of inadequate observations. When the scientific revolution, beginning in the sixteenth century, led to reliable information about the motions of heavenly objects as well as those on Earth that disagreed with Aristotle's erroneous conclusions, slavishly accepted in place of direct observations, his influence came to an end.

Exploratory Voyages

The desire to know our surroundings and to discover what the wide world around us is like manifested itself in great tours of exploration of far-away regions of the Earth. The long marches of Alexander in the 4th century BCE may be ignored for our purposes, even though he did send exotic plants and animals to his teacher and friend Aristotle for his study; Alexander meant to conquer rather than acquire knowledge.

More than a thousand years later, the sea-faring Vikings traveled all over the Atlantic, touching both northern Africa and even Constantinople to the east and Greenland to the west. But while they mostly went in order to intimidate and plunder rather than to explore and learn — the Germanic myths based on their exploits did not glorify them as seekers of knowledge but as fearsome warriors — there were exceptions. Around 1002, Leif Ericson, the Icelandic son of an outlaw, sailed in a boat with a crew of about 35 to Greenland and beyond to the coast of what is now Canada, building a small settlement in northern Newfoundland. After he adopted the Christian religion during a visit to Norway, the King of Norway sent him back with a mission to spread Christianity in Greenland and beyond. Ericson was probably the first European traveler to land on the American continent, though his aims had little to do with the acquisition of knowledge.

Perhaps the first men who could be regarded as legitimate explorers were Niccolò Polo and his brother Maffeo in the second half of the thirteenth century. Prosperous Venetian merchants, they traveled east along the ancient Silk Road to the court of the Mongol Kublai Khan, located in what is now Beijing. The great Khan received them graciously as his guests, eventually sending them back with an ambassador to the pope, requesting instruction in the Western and Christian customs. The return journey was mostly by sea. Pope Gregory X sent the requested information back via the Polos, who, along with their 17-year old son Marco and two Dominican friars, set out to return to Kublai Khan, though the two monks lost courage along the way and never made it to Mongolia. The Polos spent the next seventeen years in China, and the Khan sent young Marco on many diplomatic missions throughout his empire. After the three

Polos finally returned home, Marco wrote a book entitled *Il Milione* about his experiences, which was translated into many languages and became very popular (even though this was before the invention of movable type in Europe). Because of this book, it is only Marco Polo and neither his father nor his uncle who is generally remembered for his adventurous travel to Mongolia. As in this case, none of the many later exploratory voyages were undertaken purely as disinterested searches for knowledge; they always had an additional purpose: sometimes it was to spread a religion, sometimes to project power, but more often it was to find lucrative trade routes.

While most of us have heard of the European explorers, Columbus, Magellan, and Vasco da Gama, in the early part of the fifteenth century the explorers were Chinese. The first Ming emperors built an extensive number of ocean-going ships, the largest of which they called treasure ships. These were truly enormous vessels, dwarfing the later Spanish and Portuguese ones (See Fig. 1.1).

Figure 1.1. Zheng He's treasure ship (four hundred feet) and the *Santa Maria* of Columbus (eighty-five feet). Illustration by Jan Adkins, 1993. (From L. Levathes, *When China Ruled the Seas*, p. 21.)

The third emperor of the Ming dynasty, Zhu Di, appointed the court eunuch Zheng He as the commanding admiral of a large fleet and sent him on a number of exploratory voyages.

Zheng He's armada consisted of 317 ships, some of them treasure ships, laden with the finest silk, thin-walled porcelains, painted, glazed stoneware, and works of art. Other warships contained 27,000 or more men. They navigated by means of the stars as well as by the use of a compass needle — floating in a bed of water for easy mobility — and measured time by means of burning incense sticks. Zheng He's fleet generally sailed fairly close to the shoreline of south Asia, but it also ventured as far south of the China Sea as Timor, not far from Australia. However, they did not go east past the Philippines and across the Pacific, never coming anywhere near the American continent.

Going west, Zheng He's fleet landed in Siam and after sailing past Sumatra, visiting Burma, and rounding India they traveled as far as Hormuz at the entrance to the Persian Gulf, at the time a cultured and wealthy city, before going on to Jidda, where they learned more about the Jewish, Christian, and Islamic religions, of which China was almost entirely ignorant. The furthest west Zheng He's voyages took him was the east coast of Africa, where he sailed as far south as the Mozambique Channel near Madagascar. His mariners did go ashore in many places, exchanging goods, but they never ventured much inland. When the ruler of Ceylon gave Zheng He a less than friendly reception, he simply ordered his fleet to leave after a brief skirmish; neither war nor conquest was his purpose. But on a return voyage he avenged the insult to his emperor, sending his mariners ashore, who managed to take the unfriendly king prisoner. They brought him to the emperor's court, but although Zhu Di pardoned him, he replaced him with a friendlier ruler. (Various conflicting versions of his story loom large in Ceylonese mythology.) Other hostile encounters of Zheng He's fleet were with pirates who dared to threaten his mighty armada and were annihilated.

While the ultimate aim of these trips is not entirely clear, it was certainly in part to acquire more knowledge of the lands in the extended vicinity of the Middle Kingdom, diplomatic exchanges and trade, and no doubt to impress them with the emperor's mighty power. On the way they experienced a typhoon and witnessed for the first time the mysterious electrical phenomenon known to Western sailors as Saint Elmo's fire, a bright purplish glow appearing at the tips of a ship's mast and of other pointed objects during thunderstorms. Wherever they landed, they bartered with the natives, exchanging their own treasures for valuable goods unavailable or unknown at home, such as mahogany, an extremely hard wood very useful for replacing damaged rudders on their ships. Returning from his travels, Zheng He brought back many exotic treasures for the emperor, the most admired being some strange looking African animals never seen in China: giraffes, which they thought were the mythological beasts known as *qilin*. He even brought with him medical herbs and a number of physicians and pharmacists he had recruited on the way, who were badly needed because China experienced several deadly epidemics during the early fifteenth century. Among the many tributes elicited from foreign rulers by Zheng He's voyages there was one the emperor especially prized: China's first encounter with spectacles, which had been invented in the thirteenth century in Europe. Malaccan traders had no doubt imported them from there, and Zhu Di richly rewarded the envoy who had brought them.

When emperor Zhu Di died in 1424, his son Zhu Gaozhi succeeded him at a time when the court was in difficult financial straits. His father had fought a number of border wars and had moved the capital from Nanjing to Beijing, where he had constructed the Forbidden City at great expense, shortly thereafter devastated by fire. Short of money, Zhu Gaozhi decreed that no more treasure ships be constructed and that even the old ones not be repaired. All voyages had to be stopped immediately, and Zheng He was relieved of his command. However, Zhu Gaozhi died after only nine months in office, leaving his twenty-six old son Zhu Zhanji to became the new emperor, whose reign has been regarded by some as the peak of the Ming dynasty. Countermanding his father's decree, he sent Zheng He, by then in his sixties, on one last great voyage, his seventh. The fleet traveled as far as Mecca and Medina, but while at sea, Zheng He died and was buried in the waves following Muslim rites that they had recently learned. The armada returned to Beijing, bringing home horses, elephants, and another giraffe as tributes from Arab states. Other tributes continued to be brought to the court from many countries Zheng He's fleet had visited, until the early death of emperor Zhu Zhanji in 1435. This marked the end of China's sea voyages. Zhu Zhanji's successor, Zhu Qizhen stopped all further exploration and trade by sea, for reasons having to do primarily with court intrigues. By the end of the century even the construction of seaworthy vessels was strictly prohibited. The country that had dominated the seas and explored many coastal lands for years withdrew from any quest for further knowledge and even trade. The game was now up to the Europeans.

Not much later in the same century, Europe began its own exploratory enterprise, which we all learn about in school, beginning with Christopher Columbus. Reportedly born in Genoa, he first went to sea, by his own account, at the age of ten. Growing up as a sailor, he conceived a daring plan to travel to India, not by the usual long land route along the Silk Road like the Polos, but by sailing westward. After all, the Earth is a globe, so there should be a direct sea connection across the Atlantic that would much facilitate the lucrative trade with China and the Indies. His calculations, based on his estimate of the size of the Earth, convinced him that the distance should not be too large for a ship to carry the necessary amount of food and drinking water to sustain the crew during the long voyage. What is more, he had learned about the prevailing easterly trade winds that would help propel his ships on their Atlantic voyage westward, though of course they would be in his face on his return. In fact, he was grossly mistaken about the distance from the west coast of Europe to the coast of China and India. Owing to a variety of errors, he had vastly underestimated the true distance he would have had to travel before encountering land, and he would have been in great difficulties had he not unexpectedly bumped into the American continent.

Of course, Columbus had to find a sponsor to fund his very adventurous planned undertaking. After twice appealing to the king of Portugal and being turned down, he finally managed to convince King Ferdinand II of Aragon and Queen Isabella of Castile, the monarchs of Spain, to sponsor his voyage, with much of the financing coming from private investors. He was given the title Admiral of the Ocean Sea and, at his insistence, entitled in perpetuity to 10% of all revenues received from any lands he might discover on his way. Columbus departed in 1492 with his storied ships, the *Santa Maria* (nicknamed Gallega), *Pinta*, and *Santa Clara* (also called Niña). These were relatively small vessels, with the flagship *Santa Maria* holding a crew of 40. After five weeks they arrived at an island in what is now called the Bahamas, inhabited by people he thought he could easily conquer if so desired. After exploring the east coast of Cuba, the *Santa Maria* had to be abandoned because it had run aground, and Columbus returned to Spain with his other two ships, leaving a small settlement behind in Haiti and kidnapping a few natives to take home with him. His crew however brought back another souvenir of their voyage, *syphilis*, which rapidly spread across Europe with devastating effect after many of the crew served in the army of King Charles VIII.

Columbus led three more voyages across the Atlantic, searching in vain for a passage further west towards India and China, even though by that time the Portuguese explorer Vasco da Gama had already discovered another direct sea route there, going south around Africa via the Cape of Good Hope. On Columbus's third trip he made his first landing on the South American mainland near Trinidad and ventured to explore the Orinoco river. Once, during a hostile encounter with natives in his fourth voyage, he was saved by the superior European knowledge of astronomy. His successful prediction — using an ephemeris constructed by the astronomer Regiomontanus — of the lunar eclipse observed on 29 June, 1504, impressed and intimidated them so much that they provisioned his hungry men until help arrived and he was able to leave with his crew. Like Thales more than two thousand years before, Columbus discovered not only that such awesome heavenly phenomena as eclipses could have important psychological effects, but that to have sufficient knowledge to be able to predict their occurrence was regarded as magic.

Supported by the king of Portugal, the exploratory voyages of Vasco da Gama began in 1497, with a fleet of three ships and a crew of 170. After a record-breaking sailing trip of more than three months and 6000 miles of open ocean without sight of land, they crossed the equator and sailed around the Cape of Good Hope, taking advantage of the south Atlantic westerlies, and landed on the east coast of Africa. Continuing on, the fleet at times supported itself by piracy, looting unarmed Arab merchant ships, before arriving near Calicut in India in May 1498, where the Portuguese encountered some resistance from the local ruler and indigenous Arab traders. The return trip, against the prevailing winds, was very slow and beset by scurvy, and only two of his ships made it back to Portugal. However, the newly appointed 'Admiral of the Indian Seas' and his homeland had learned much from his exploration, and the Portuguese economy reaped profits for a long time. He made two further voyages along the same route with much larger fleets of warships, but their purpose was clearly the forceful projection of Portuguese power rather than the acquisition of knowledge.

At approximately the same time as da Gama's third voyage to India around Africa, another Portuguese explorer was funded by the Spanish King Charles V to renew Columbus's unsuccessful attempt to find a westward route there. Ferdinand Magellan nursed an ambition even larger than that of Columbus, namely to circumnavigate the

Earth for the first time. He set out in 1519 with a fleet of five ships containing a crew of 237 men, knowing, of course, from the experience of Columbus that the American continent was in his way. Sailing south he found a passage through the tip of South-America north of Tierra del Fuego via the strait that now carries his name, his fleet reaching the Pacific Ocean and succeeding in his goal of sailing around the globe. Magellan himself however did not see his great purpose realized. Shot by a poisonous arrow during a skirmish with native forces on the Philippine island Mactan, he died before his voyage had closed the circle.

If Columbus, da Gama, and Magellan could be regarded as genuine explorers, Columbus's discovery of the American continent also opened up its riches for plunder, especially after vast stocks of gold were seen there by subsequent travelers. The Spanish Court made full use of this new treasure house by sending the conquistadors Hernán Cortés and Francisco Pizarro, the first to Cuba and Mexico and the second to Panama and on to Peru. Owing to the superior European technology — firearms and gun powder, the possession of horses, which had not been domesticated by the native people, and most of all, their immunity to diseases they brought with them which decimated the natives, the conquistadors rapidly managed to plunder great hoards of gold from the empires of the Incas and the Aztecs as well as silver mines in Mexico to enrich the Spanish Crown. At the same time their conquest destroyed the highly developed native civilizations. These were the unfortunate after-effects of searching for knowledge; if knowledge is power, that power was not always used for good.

In the eighteenth century, the age of Enlightenment, geographical exploration was taken up again seriously by Europeans. Joseph Banks, who later served as the president of the Royal Society for many years, made his name as a young man when he accompanied Captain Cook on the *Endeavour* for an expedition to sail around the world with the purpose of making astronomical observations from Tahiti, to explore the southern regions of the Pacific Ocean — searching for the possibility of another continent south of Australia before the French could find and colonize it — and to collect biological and botanical specimens there. Banks signed on as the official botanist — there was also an astronomer on board — intending to make and record detailed botanical and biological observations wherever they landed in addition to the fish, the birds, and the natural phenomena he encountered on the way. Owing to the success of their experimental diet supplemented by sauerkraut and fresh bird meat to avoid the dreaded scurvy, as well as Captain Cook's brilliant seamanship, the *Endeavour* landed on the island of Tahiti early for the transit of Venus they were sent to observe, necessitating an extended stay. As a result Banks spent a few months there, living with the natives, learning their language, studying and meticulously recording their practices — as well as greatly enjoying some of their exotic sexual customs in what he regarded as an earthly paradise — eventually even bringing one man back to London with him as a 'specimen.'

The sea expeditions of the nineteenth century were mostly concerned with the areas near the north and south poles. In 1818, the Scottish naval Commander John Ross (later knighted, Sir John) was sent on an unsuccessful voyage with two sail ships to find the long-sought Northwest Passage that would lead from the north-Atlantic to the Bering Strait through the Arctic Ocean around the American continent. (This passage was eventually first navigated by the Norwegian explorer Roald Amundsen on a voyage that took from 1903 until 1906.) Eleven years later he made another attempt with a side-wheeled steam vessel, which took him to a hitherto unexplored area past Lancaster Sound, where the ship became stuck in an icepack for four years. Using this period for further explorations, Ross found the magnetic north pole, which had been known from compass-navigation to differ from that of the geometric pole but whose precise location — which changes somewhat in the course of time — had been unknown. Ross and his crew were finally picked up and brought home by a British vessel.

In 1839 Sir John Ross's nephew, Sir James Clark Ross, who had accompanied his uncle on his northern expeditions, organized his own expedition to the continent of Antarctica on two 'bomb vessels,' warships with especially strong hulls, which proved themselves to be very valuable in withstanding the pressure of thick ice. Touching Australia and New Zealand, they circumnavigated Antarctica along the South Pacific, discovering what came to be named the Ross Sea and Victoria Land. They sailed as well near two volcanoes on the edge of the continent and around what was later called the Ross Ice Shelf but made no attempt to explore the interior of Antarctica.

There had, however, been a much earlier expedition to the north, for an entirely different purpose from that of Sir John Ross. During the eighteenth century, the French Academy had been roiled by a geodesic controversy: if the Earth is not precisely spherical, is its shape prolate, i.e., an oblong spheroid, with the distance between its poles greater than its diameter at the equator, or is it oblate, more like a pancake, flattened at the poles? Descartes had decreed the first, while Isaac Newton the second, and the mathematician Pierre Louis Moreau de Maupertuis had shown that, irrespective of the nature of the force of gravity — about which Newton and Descartes disagreed — any deformable rotating sphere would become flattened at its poles. He intended to find out if the Earth confirmed his theory by making precise measurements near the North Pole of the distance along a meridian — a north-south line on the Earth — corresponding to one degree of latitude, as measured by the fixed stars. If the Earth were oblate, this distance should be longer in the north than near the equator; if prolate, it should be shorter — if it were exactly spherical, of course, this distance should not vary with the latitude. The question of the Earth's shape could be answered most easily by precise measurements of the period of a pendulum at various latitudes: oblateness would imply that the pendulum is closer to the center of the Earth near the poles than at the equator and so experience stronger gravity, whereas prolateness would imply it is weaker near the poles. But then the conclusion would depend on Newton's theory of gravity, which in France was still under contention. So Maupertuis meant to travel north, measure the meridial distance corresponding to one degree of latitude and compare it both to earlier measurements along the Paris meridian further south and to other measurements made by an expedition to Peru in 1735.

After obtaining funding from the King of France, Maupertuis left in 1737 on an expedition 'to the North Pole,' taking along all the needed astronomical and land-surveying instruments. There was, of course, no need to go as far as the North Pole; it was difficult enough to lug all the required apparatus to northern Lapland, beyond the arctic circle, where he actually performed the intended measurements. The result confirmed his theory: the Earth is indeed oblate, like a pancake. Here was an expedition that had actually been undertaken for the sole purpose of acquiring knowledge (except perhaps also to enhance Maupertuis's reputation). The only other comparably 'pure' expeditions were those of the astronomer Sir Arthur Eddington to the island Principe near Africa to observe the solar eclipse of May 29, 1919 in order to measure precisely the bending of starlight by the gravity of the sun as predicted by Einstein's new General Theory of Relativity — a verification that instantly made Einstein world famous — and Darwin's long voyage onboard the *Beagle*, whose profound after-effects were destined to last to this day.

Subsequent geographical exploration turned to the largely mysterious jungles of sub-Saharan Africa. All the first expeditions sent there ended tragically, the travelers never to return. The first successful exploration was conducted at the end of the eighteenth century by a Scottish romantic named Mungo Park, who spent more than two years in the jungle with a small group of companions and, upon his return to London, wrote a book entitled *Travels in the Interior of Africa*, which became an instant best-seller. However, when he made a second expedition to west Africa some eight years later, funded by the Colonial Office, he did it in force, accompanied by a group of armed soldiers. From this, Mungo Park never returned.

The best known of the later African explorations were, of course, those led by David Livingstone and Morton Stanley. Livingstone was a Scottish missionary and a physician, whose first trips to southern Africa were strictly for the purpose of spreading Christianity. Becoming frustrated in this endeavor, he turned to exploration in 1854. Accompanied by a small group of porters and only lightly armed for protection, he traversed the continent from Luanda on the Atlantic to the mouth of the Zambezi river on the Indian Ocean, 'discovering' for non-African eyes the great waterfalls he named after Queen Victoria. Returning to Britain he was celebrated as a great explorer and resigned from the Missionary Society, which was pressing him to continue his religious work. In 1858 he went back to Africa for an ambitious attempt to open up the river Zambezi for navigation, at which he failed because of cataracts and impassible rapids; the expedition was terminated in 1864. However, he returned again two years later to search for the source of the river Nile, at which he was also unsuccessful. Ill and lost, he had no contact with the outside world for six years until he was found in 1871 by Henry Morton Stanley.

Stanley was a Welsh journalist, hired by the *New York Herald* after a number of adventurous exploits, including an expedition to the Ottoman Empire. In 1869 his employer sent him to Africa with the explicit goal of finding David Livingstone, the explorer lost there in the jungle. In 1871 he traveled to Zanzibar and set out into the wilderness, well equipped and outfitted, accompanied by some 200 porters, whom he treated brutally. After experiencing great difficulties in the jungle, with diseases as well as desertions, he finally found Livingstone near Lake Tanganyika, claiming to have greeted him with the subsequently famous words 'Dr. Livingstone, I presume?'

in accounts of the meeting in the *New York Times* and the *Herald*. Livingstone and Stanley then went on exploring the nearby region of Africa.

In 1874, Stanley returned to Africa for another expedition, leading 356 people with the purpose of tracing the course of the river Congo. After almost three years of enormous physical challenges as well as diseases and hostile natives he arrived at the mouth of the river, all but 114 of his companions dead. His book *Through the Dark Continent* describes his tribulations. Stanley's later African expeditions helped establish the Congo as the property of the Belgian King Leopold — a colony that became notorious for its brutal treatment of the natives — and tarnished Stanley's name for alleged extreme cruelties. European exploration of the interior of Africa was beset by great difficulties that had to be overcome heroically to gain knowledge of the mysterious continent, but the wounds left would bleed for a long time.

Other famous explorations in the early twentieth century were led by Robert Peary and Frederick Cook to the North Pole and by Roald Amundsen to the South Pole, which Ernest Shackleton had almost reached before having to return. Cook and Peary reported reaching the North Pole after great hardships, the former in 1908 and the latter in 1909, but controversy surrounded their claims. However, in contrast to the much earlier expedition by Maupertuis to Lapland and the expeditions by Sir John Ross and Sir James Clark Ross, these voyages, like the climbing of the highest mountains, should not really be regarded as motivated by a search for knowledge; they were done partly for the sake of conquest but most of all to demonstrate — successfully — man's capacity for endurance. The harder a place on the Earth was to reach, the more challenging and attractive to the adventurous simply because it was there.

Exploration of the Earth was not restricted to the land surface. Measurement of the depth of the ocean and sampling of its floor sediments had already been done by the Vikings using sounding weights consisting of a lead weight attached to a line. In the course of time, these kinds of measurements and samplings became increasingly sophisticated. By the middle of the twentieth century technology had sufficiently advanced to construct vessels allowing humans to descend and observe the nature of the deep ocean, including its flora and fauna. The American explorer Charles William Beebe designed a spherical steel vessel, a 'bathysphere,' in which he descended in 1930 to great depths and described by cable telephone what he saw through a porthole to his collaborator Otis Barton. By 1948, the Swiss physicist August Piccard had invented and tested a much sturdier vessel he called a 'bathyscaphe,' in which he reached a depth of 4,000 meters. Eventually, his son Jacques Piccard and Navy Lieutenant Donald Walsh descended in an improved version of the original bathyscaphe called *Trieste* (see Fig. 1.2) to a depth of 10,915 meters at the lowest known point on Earth, the Challenger Deep in the Mariana Trench in the Pacific Ocean.

Figure 1.2. The bathyscaphe *Trieste* lifted out of the water, circa 1958–59. (US Naval Historical Center Photograph, from geology.com.)

In the meantime, the search for knowledge had already reached beyond the Earth. After Galileo had turned the newly invented telescope to the heavens, this new instrument, vastly more powerful than the naked eye that had been employed for millennia to observe the moon and the stars, yielded enormous amounts of fascinating new information, some of it contaminated by fantasies. Sir William Herschel, assisted by his younger sister Caroline, built his own reflecting telescopes (a design invented by Isaac Newton that enabled much shorter and less cumbersome instruments to produce as much magnification and light-gathering power as the older refracting kind employed by Galileo (see Fig. 1.3)) and used them to study the surface of the Moon. Perceiving much more detail than had been observed before, he imagined he saw forests and was convinced there was life there. It was not until the twentieth century that the exploration of the

Figure 1.3. The forty-foot reflecting telescope that Sir William Herschel built for himself, with the help of his sister. (From Holmes, *The Age of Wonder*, p. facing page 141.)

Moon could go beyond telescopic viewing when in 1969, after years of preparation, the rocket ship of the American *Apollo 11* mission landed on our satellite and Commander Neil Armstrong became the first human to set foot on the Moon, uttering the words that would echo around the world: 'That's one small step for a man, one giant leap for mankind.'

No telescope was powerful enough to see as much detail on the planet Mars as could be seen on the Moon, so astronomers' imagination had free reign. After the Italian priest Pietro Secchi in 1876 thought he could make out some details he called 'canali,' even the well known astronomer Percival Lowell was convinced he could see irrigation 'canals' on the red planet that he thought must have been constructed by intelligent creatures. While no human beings have yet been able to visit Mars, robot vehicles sent there by rockets since the 1960s by various nations have been able to explore its surface composition sufficiently to make the presence of life there extremely unlikely, though the question is still open.

It is clear that unless the search for knowledge is disciplined by rational control, it can be led wildly astray by imagination and prior beliefs. Let us turn then to the development of science, the very instrument to exert such rational control.

2

We Want to Understand

One of the legacies the ancient Greek civilization bestowed upon us was the systematic study of nature for its own sake, simply to satisfy our urge to understand it. Think of the philosopher Democritus of the fifth century BCE, who regarded certain hard, solid, invisible, small particles that differed from each other only in shape and arrangements — his atoms — as the ultimate constituents of matter. Or think of Archimedes of the third century BCE, arguably the first person who could be called a real scientist in the modern sense, and his law of the buoyancy of floating bodies. That is not to say, of course, that Greek science had no utilitarian or ritualistic motivations. Ptolemean astronomy — Ptolemy was an Egyptian, but his thinking was entirely Aristotelian — was still driven at least in part by the desire to predict eclipses for religious purposes, Pythagorean mathematics had deeply mystical motivations, and Archimedes applied his science directly to military purposes. There is, however, little doubt that the more than casual attempt to look into the inner workings of nature, purely for the sake of understanding, began with the Greeks. The same is true of mathematics. While the beginnings of geometry go back to Babylonian and Egyptian methods of land-surveillance, both the systematization of geometry and the study of numbers were first undertaken by the ancient Greeks.

The origins of what we would now call 'pure science' and 'pure mathematics' thus go back about twenty-five centuries. Afterwards, however, both pure science and pure mathematics languished for a very long time; though they were not forgotten, they remained frozen for about seventeen hundred years. The only significant exception was the advancement of algebra by the Arabs. Although there were a number of important technological innovations during this period of time, particularly in China, many of them later duplicated in Europe, there was almost no progress in basic science.

Modern science, in the sense in which we now use that term, began no earlier than the sixteenth century, though much was foreshadowed by the alchemists. Relying mostly on magical formulas and incantations, their aim was primarily the transmutation of elements into one another, especially into gold. (The legendary character Doctor Faustus, entering classical literature via Christopher Marlowe and Janhann Wolfgang von Goethe, was an alchemist who sought to acquire universal knowledge and the resulting power by making a pact with the devil.) Nevertheless, the alchemists began to recognize the atomic constitution of matter. It was the achievement of Robert Boyle in the seventeenth century to change alchemy into what we now call the science of chemistry. (However, he never managed to convince Isaac Newton to quite abandon his belief in alchemy.)

The rise of modern science during the last three hundred years has been truly spectacular. Not only has there been an enormous increase in our knowledge, but there has been an equally impressive growth in the number of people who devote a large part of their lives to enhancing that knowledge. For example The American Association for the Advancement of Science, which was founded in 1848 with 461 members, now has a membership of over one-hundred and thirty-thousand, and it is an often-repeated fact that more than half the scientists who ever lived during the entire history of the world are alive today.

What were the fundamental aims and purposes of the scientists and mathematicians who contributed to this explosive expansion of knowledge? I invite you to compare the work of an unquestionable genius such as Leonardo da Vinci with that of Galileo. In addition to being an unsurpassed painter, Leonardo was a most ingenious inventor of technical devices, and he offered the valuable services of his technical imagination and ingenuity to dukes and princes to enhance their military power. Even though medical science undoubtedly benefited from his drawings and studies of the insides of the human body, we do not generally consider him a scientist. On the other hand, although Galileo's telescope at first served, for more than any other purpose, as a very useful tool for navigation, such an application was not his primary purpose, and he is regarded as the modern scientist par excellence.

There are two principal reasons why modern science is considered to have originated with Galileo Galilei and Isaac Newton. One is that they based their understanding of nature on observation and experiment and did not believe that such understanding could be gained by pure thought alone, as Plato and many of his medieval followers believed. Ground breaking experiments in the seventeenth century by Robert Boyle with his air pump, which demonstrated the falseness of Aristotle's doctrine that 'nature abhors a vacuum,' encountered much opposition from philosophers as a matter of principle because they considered the vacuum a philosophical concept not subject to experimental test. While they did not accept the many erroneous conclusions Aristotle had drawn from his relatively primitive observations, Boyle as well as Galileo and Newton were still followers of the ancient philosopher's basic teachings that knowledge and understanding of nature could be gained only through experimentation and observation.

The other reason why modern science is regarded as having its beginning with Galileo and Newton is that their quest for knowledge was not driven by any desire for useful applications. Not that they were hostile or even indifferent to such applications; they were neither. But their basic motivation was not to seek new knowledge for the benefit of society, or to enhance the power of their nation, king, or duke. It was to understand the world of nature. For any observed phenomenon they wanted to know both the how and the why — why in the sense of what its underlying cause was rather than in the teleological sense of to what end. When Gregor Mendel, a priest in an Augustinian monastery, asked himself why the hybrids of green and yellow garden peas produced three times as many yellow peas as green ones, he was not asking why God produced this exact ratio he had painstakingly observed, or what purpose it served; he asked what its explanation was.

The three hundred years since the publication of Newton's *Principia* has seen not only a great growth in our knowledge of nature but also a large advance in technical know-how in all areas of life. In those regions of the world that have direct access to these technical skills, the standard of living, the health, and the lifespan of the vast majority of the population have greatly increased. No one will claim that there is no connection between these two developments. In fact, almost everyone credits the advances in science with the social benisons of the past three centuries. That there have also been great costs, such as the painful adjustments caused by social changes engendered by innovations and the increased destructiveness of wars, is undoubtedly true. Sometimes these costs are held against science to an extent that outweighs acknowledgement of its great benefits, but that is beside the point. Without the growth of modern science since the Renaissance, Western society would be much the poorer.

As a result of the correctly perceived correlation between advances in science and social good many people mistakenly think one can enhance this correlation by supporting and encouraging primarily those scientists whose motivation is not the disinterested, pure desire to understand nature but who are driven by the wish to benefit mankind and therefore perform research that is relatively close to applicability. However, useful technical innovations can be created in abundance only within a matrix of general scientific knowledge; without it inventive thinking remains sterile.

It is generally acknowledged that a man like Thomas Edison, who was not a scientist but an inventor, made many important technical contributions that were extremely beneficial to society at large. But such technological advances occur only rarely and slowly without an ambient background of basic understanding of the world as their underlying foundation. Surely it is no accident that the number of technological advances in Europe before the Renaissance, or in China, is utterly negligible when compared to the number during and after the eighteenth century in Europe. Granted, much of the important work of Laplace and Lagrange in celestial mechanics was motivated to some extent by their desire to determine if the solar system was stable or whether there was some danger that humanity's life might be cut short by a celestial catastrophe. And very few scientists or mathematicians, especially in the nineteenth century, adopted the extreme attitude of the modern English mathematician G. H. Hardy, who proclaimed himself proud never to have worked in an area of mathematics that had the slightest chance of ever being useful or to have any applications. (He turned out to have been wrong in this. The theory of numbers, which was his main field of work and to which he made many important contributions, was later found to be extremely useful to cryptography.) In fact, during the nineteenth century there did not exist the same kind of dichotomy between pure and applied science, or between pure and applied mathematics, as we have today. Nevertheless, the great scientists Galileo, Newton, Lavoisier, Boyle, Lagrange, Laplace, Darwin, Pasteur, Helmholtz, Kelvin, Boltzmann, Gauss, Faraday, Maxwell, Gibbs, Curie, Planck, Einstein, Lorentz, Poincaré, Bohr, Rutherford, Heisenberg, Schrödinger, Dirac, Muller, Fermi, and many others too numerous to mention, were not motivated primarily by the desire to find new ways to be useful to the world. They wanted to understand what made nature tick. As Einstein — who was not a religious man but often liked to use God as a metaphor — put it:

> I want to know how God created this world. I am not interested in this or that phenomenon. I want to know His thoughts; the rest are details.

The science of medicine clearly constitutes an exception to this argument. The motivation of the medical scientist is and has to be largely the good of humanity. However, even in this field it is well to remember the words of Hippocrates: 'The nature of the body is the beginning of medical science.' Today, of course, we include in the nature of the body its microscopic constituents, discovered and elucidated purely for the sake of understanding. Louis Pasteur was perhaps the greatest exemplar of achievements in this area.

The first thing to be clear about science is that it does not consist of a simple collection of facts assembled by observation. Scientists are not clones of Thomas Gradgrind in Dickens's *Hard Times*, who expostulated,

> "Fact, fact, fact!" and "fact, fact, fact! . . . You are to be in all things regulated and governed by fact. We hope to have, before long, a board of fact, composed of commissioners of fact, who will force the people to be a people of fact, and of nothing but fact. You must discard the word Fancy altogether. You have nothing to do with it."

Collecting facts by observation and experimentation is the first step, but the aim of scientists is to find as many connections between these facts as possible and to understand and explain them. For some of the creators of the scientific edifice, the urge to understand and decodify the universe around us had an aesthetic motivation; for others it had a mystical, perhaps even a religious, component. But their motivation was not primarily to do something useful (medical scientists excepted), nor did they simply collect data and deduce explanatory theories from them. Consider what John Maynard Keynes, the English economist, who collected Isaac Newton's unpublished manuscripts as a hobby, said about the great scientist:

> In the eighteenth century and since Newton came to be thought of as the first and greatest of the modern age of scientists, a rationalist, one who taught us to think on the lines of cold and untinctured reason. I do not see him in this light. I do not think any one who has pored over the contents of the box which he packed up when he finally left Cambridge in 1696, and which, though partly dispersed, have come down to us, can see him like that. Newton was not the first of the age of reason. He was the last of the magicians, the last of the Babylonians and Sumerians, the last great mind which looked out on the visible and intellectual world with the same eyes as those who began to build our intellectual inheritance rather less than 10,000 years ago. Isaac Newton, a posthumous child born on Christmas Day, 1642, was the last wonder-child to whom the Magi could do sincere and appropriate homage.

If this strikes you as a bit of romantic hyperbole, I agree. But what is crucial is that the essential business of scientists is to explain the observed world by constructing theories. And, contrary to what some philosophers of science used to believe, theories are not derived from given facts by means of a mechanical process called induction. They are produced by what Thomas Gradgrind sneered at as Fancy, but are shored up and confirmed by experiments and observations.

When Isaac Newton made his famous disclaimer 'I do not make hypotheses' he meant that he did not indulge in speculation, as was sometimes customary among natural philosophers and scientists at the time. His gravitational theory was grounded in observed evidence. Nevertheless, such evidence does not lead to a general theory by a purely logical process; what is required, as Einstein put it, is 'intuition supported by rapport with experience.' Indeed, the axioms that form the basis of fundamental physical theories, Einstein averred in his Spencer Lecture in 1933, are 'free inventions of the human intellect.' He did not mean to deny that the 'free inventions' must finally be anchored in experimental observations, but they are not determined by them. There is an enormous gap between falling apples, rolling balls, and planetary orbits on the one hand, and the equations of motion and the law of universal gravitational attraction on the other; 'there is no logical bridge from experience to the basic principles of theory' (Einstein). In addition to the original inference that leads from 'rapport with experience' to a scientific theory, there is the further constraint that predictions of the theory must be verified by subsequent observation. (The word prediction in this context does not necessarily refer to an observation in the future but it may mean the elucidation of an already known experimental or observational fact that was not taken into account in the formation of the theory.) 'The process by which we come to form a hypothesis is not illogical but nonlogical, i.e., outside logic,' wrote the biologist Peter Medawar, 'but once we have formed an opinion we can expose it to criticism, usually by experimentation.'

In many instances the driving motivation of theoretical scientists, and of mathematicians as well, is so far removed from ideas of applicability that it is closer to that of an artist. While for some there is a mystical awe of nature that approaches a religious feeling, for many others it is an admiration of beauty. To be sure, the beauty of a mathematical structure, of an equation, or of a scientific theory can be appreciated only after the beholder has acquired the needed kind of training and knowledge, just as some other greatly–admired kinds of beauty require a highly educated taste to be fully appreciated. On the occasion of the death of the mathematician Emmy Noether, Einstein wrote in the *New York Times*:

> Pure mathematics is, in a way, the poetry of logical ideas.
> One seeks the most general ideas of operation which will

bring together in simple, logical, and unified form the largest possible circle of formal relationships. In this effort toward logical beauty spiritual formulas are discovered necessary for the deeper penetration into the laws of nature.

There is, however, one essential difference between the aesthetic motivation of an artist and that of a scientist: while the artist is subject to no other authority, the scientist has to bow before the final arbiter of agreement with facts as revealed by experiment or observation. Between a scientist's soaring imagination and the need to discipline that free flight of thought by critical comparison with experimental data there is a continual tension. 'Let the imagination go, guiding it by judgment and principle,' advised Michael Faraday, the great nineteenth-century experimental scientist, 'but holding it in and directing it by experiment.' This aspect of the scientist's creation clearly distinguishes him from the artist, who is not subject to such constraint. And it is this aspect, Ernest Rutherford emphasized, which assures us that 'the physicists have ... some justification for the faith that they are building on the solid rock of fact, and not, as we are often so solemnly warned by some of our scientific brethren, on the shifting sands of imaginative hypothesis.'

Nevertheless, aesthetics plays an important role both in the initial plausibility of an announced result and in the value it is accorded after its validity has been accepted. Not only do scientists accept beautiful theories more readily than ugly ones, but they also greatly admire beautiful and elegant demonstrations and put a lower value on clumsy and unnecessarily complicated ones. 'What I remember most clearly,' the astrophysicist H. Bondi wrote, 'was that when I put down a suggestion that seemed to me cogent and reasonable, Einstein did not in the least contest this, but he only said, "Oh, how ugly.".... He was quite convinced that beauty was a guiding principle in the search for important results in theoretical physics.'

Similarly for mathematicians, who also greatly value beauty and elegance: no matter how beautiful a purported theorem is, it has to be logically correct, and proved to be so, in order to be accepted as a theorem. However, aesthetics plays an important role both in the initial plausibility of an announced result and in the value it is accorded after its proof has been accepted. Even though the great Indian mathematician Srinivasa Ramanujan, who grew up without the benefit of extensive education in traditional mathematics, announced many astonishing mathematical propositions without proof, his English mentor G. H. Hardy and other mathematicians were almost immediately convinced of their truth in part because of their beauty. This, of course, did not obviate the need for actually proving them before they could be accepted as theorems (which sometimes turned out to be very difficult), but although a few were found to be incorrect, beauty contributed to the *prima facie* case for truth. Not only do mathematicians accept beautiful results more readily than ugly ones, but they also greatly admire beautiful and elegant proofs and put a lower value on clumsy and unnecessarily complicated ones, even when they are reluctantly forced to accept them as correct.

Apart from the motivation of scientists, what distinguishes science from all other ways of acquiring knowledge is, first and foremost, reliability. When we believe information transmitted to us in some form, as originating by tradition or authority, by our senses, or by pure thinking alone, the question that nags us always is whether it might not turn out to be illusory. How can we be sure? If it is based on tradition or authority, there is no way of ascertaining its validity without questioning the originating tradition or authority, a method that is usually socially unacceptable. The mere act of doubting will be rejected and the questioner may be ostracized. Such knowledge therefore discourages the very act of its improvement and advancement. What is more, if knowledge based on one source of scripture is not accepted, what should take its place: belief in an alternative authority? Disagreements between believers in different authorities have led to many bloody wars in the course of history. Knowledge originating from the senses is certainly not infallible, but errors can be corrected by new observations, and those again are subject to possible revision by further investigation. Disagreements between scientists, though certainly not uncommon and sometimes fierce, have never led to armed clashes. During the nineteen eighties, strong disagreements among natural and social scientists about the objective validity of scientific knowledge were known as the 'science wars' (see, for example, my book *The Truth of Science*), but the name was, of course, a metaphor. Knowledge based on pure thought, as in mathematics, is subject to correction by further thought and ultimately to proof. (I will not treat the deep philosophical question of the nature of mathematical proof in this book, nor shall I further pursue the origin of mathematical knowledge.)

Some philosophers of science used to construct a variety of rigid procedures they regarded as 'the scientific method.' Such attempts have always failed. The scientific method — without quotation marks — was well described by the biologist Thomas Huxley (one of the strong defenders of Charles Darwin) as 'simply common sense at its best; that is, rigidly accurate in observation and merciless to fallacy in logic' and more pithily in the words of the American physicist Percy Bridgman, 'to use your noodle, and no holds barred.' Nevertheless, there are, of course, certain crucial touchstones. A given theory may have been suggested to its originator by some observational or experimental data, but it has to imply some new data that are again subject to observational or experimental tests. Without such predictions a theory is worthless. However, no amount of corroboration can ever establish the correctness of a theory forever. Since instances of testing occur without end, a scientific theory is always provisional.

Newton's laws of motion and his law of universal gravitation were enormous achievements, verified by a myriad of tests. Nevertheless, three hundred years later, they were found to be correct only within a certain area of applicability, the only observations to which Newton had access, namely for objects that move slowly compared to the speed of light and whose masses are not too large. In a more general context they have to be replaced by new, more general laws later introduced by Albert Einstein. What is more, at the scale of atoms — a scale that too was not yet accessible to Newton — the entire framework of Newton's mechanics had to be replaced by that of quantum mechanics. For all we know, both Einstein's theory of relativity and quantum mechanics will have to be replaced again at some future time.

In order to get some feeling for the way scientists go about their work, let us look at some concrete examples of great scientists and their major achievements in various fields: Charles Darwin, Gregor Mendel, Louis Pasteur, Michael Faraday (all lived during the nineteenth century), Max Planck half in the nineteenth and half in the twentieth, and Enrico Fermi in the twentieth. (Another exellent example is, of course, Marie Curie, but her story is so well known that I will not retell it here.)

Charles Darwin

The extremely able, twenty-six-year old Captain Robert FitzRoy commanding the *Beagle* was preparing to undertake a voyage to survey parts of the Southern Hemisphere, including South America. He was looking for a man who could not only do the needed detailed observations of the flora and fauna they encountered on the way but who could also serve as his companion with whom he could have interesting conversations during the long trip. The young man who presented himself was Charles Darwin, an enthusiastic naturalist just graduated from Cambridge University, especially after reading Alexander von Humboldt's *Personal Narrative of a Journey to the Equinoctial Regions of the New Continent*. After pointing out to the eager but inexperienced candidate the inconveniences and

Figure 2.1. Water-color portrait of Charles Darwin in late 1830s, by George Richmond.

hardships to be expected on the planned extended voyage, FitzRoy was sufficiently impressed to accept him, and after some delay the *Beagle* finally left Plymouth in December 1831.

Two months later, the ship landed on the coast of what is now Argentina, where they wandered by foot and horseback through the rain forest. Darwin began collecting birds and plants on land as well as marine animals in a net trawling behind the ship. He even hacked out of coastal hillsides fossils they discovered, which he recognized as the remains of extinct animals and which to his great surprise seemed to resemble closely some of the creatures now living nearby. Wherever he could he sent his mounting collection of specimens and fossils back to Cambridge by the naval postage system, and he meticulously recorded and catalogued everything with all the scientific exactness he had picked up at Cambridge. His reading of the recently published first volume of Charles Lyell's *Principles of Geology*, which FitzRoy had presented to him upon boarding, had inflamed Darwin's scientific ambition and changed his way of looking at the history of the Earth. (The second volume of Lyell's book reached him later on the voyage.) As for being a congenial conversation partner to FitzRoy, things did not go so smoothly. The Captain turned out to be a very conservative fundamentalist Christian, and though the two became good friends, they had many intellectual clashes. What is more, Darwin was prone to seasickness, which did not help conversationally.

By the middle of 1834 they reached the Pacific Ocean after passing through the Magellan Strait at Tierra del Fuego and began to turn northward along the west coast of South America. There they saw from off shore the eruption of a nearby volcano, followed by a severe earthquake as well as a tsunami. Long exploratory horse rides after landing allowed Darwin to see the devastating after-effects of these events and to collect more specimens of animals and vegetation as well as fossils. From there the *Beagle* sailed six hundred miles west of Ecuador to the Galapagos Islands, where the strange volcanic landscape and the unflinching behavior of beasts and birds had been reported by previous visitors. Amid surroundings that reminded him of Milton's description of hell in *Paradise Lost* — a copy of which he carried with him — Darwin was astounded by the strangeness of the animals he encountered. Penguins in the tropics? And varieties of weird, big lizards — iguanas — pervaded the place. He collected multitudes of specimens both of the strange fauna and the large number of unusual plants. In many cases these collections, which seemed to him at the time duplicates of one another, turned out, upon closer examination at home, to be different species.

From the Galapagos Islands, the *Beagle* sailed on via Tahiti to New Zealand and Australia. Though repelled in Sydney by the behavior he observed of the newly rich descendants of former settlers who had originally been taken there as convicts, Darwin was fascinated by such strange animals as the platypus. Some of the birds, such as parrots, also seemed unusual, whereas others, like crows and magpies, appeared to closely resemble their English counterparts. Sailing further west into the Indian Ocean moved on to via the Cocos (Keeling) Islands, they saw coral reefs up close for the first time, moved on to Mauritius, stopped at Cape Town for a short visit, and then northward back home to England. The *Beagle* landed at Falmouth on October 2nd 1836. They had been away for five years, circumnavigating the globe.

The long sailing trip had been of a purely exploratory nature, not undertaken to gather evidence for the theory of evolution, which Darwin originated only after returning. Along the way he had been learning about the evidence for the changing nature of the Earth in the course of its long history by reading the two volumes of Charles Lyell's *Principles of Geology*, and on his return he read the recently published *Essay on the Principle of Population* by Thomas Malthus, which impressed upon him the constant struggle for survival by all living beings. With these ideas in mind, Darwin contemplated the flora and fauna he had seen and collected on his voyage with captain Fitzroy, coming to the conclusion that not only were species evolving in the course of time rather than being fixed once and for all, but that the mechanism driving this evolution was what he called natural selection. Here is how he put it in his great book, *On the Origin of Species by Means of Natural Selection* (page 6):

> Although much remains obscure, and will remain obscure, I can entertain no doubt, after the most deliberate study and dispassionate judgment of which I am capable, that the view most naturalists entertain, and which I formerly entertained — namely that each species has been independently created — is erroneous. I am fully convinced that species are not immutable; but that those belonging to what are called the same genera are lineal descendants of some other and generally extinct species, in the same manner as the acknowledged varieties of any one species

39

are the descendants of that species. Furthermore, I am
convinced that Natural Selection has been the main but
not exclusive means of modification.

While each generation inherited the characteristics of the previous generation — he knew of course neither the laws governing
heredity nor its biological mechanism, as they were not discovered
until later — this transmission of characteristics was never perfect but
always subject to small variations of unknown origin. As a result, each
new generation of a species contained a mixture of individuals, not
all of which were equally well adapted to their surroundings, and
those best adapted won out over the others by producing more
offspring, eventually crowding out the less adapted and surviving
as a changed species. As conditions on the Earth changed over a
long course of time, species changed as well because different sets
of characteristics proved themselves better adapted to the altered
surroundings. This is what he meant by the term *natural selection*,
and it came to be enormously controversial. It was attacked because it
offended those who either believed that species were created by God
and immutable — a belief that Darwin shared when he set out on
his voyage — or else that if there was any evolution, its remarkable
results had to be guided by an intelligent designer. Darwin was fully
conscious of such criticisms (which are raised by some to this day)
and he met them on page 2:

> . . . I am well aware that scarcely a single point is discussed
> in this volume on which facts cannot be adduced, often
> apparently leading to conclusions directly opposite to
> those at which I have arrived. A fair result can be obtained
> only by fully stating and balancing the facts and arguments
> on both sides of each question[.]

As he was religious himself, and his wife more so, he was extremely
hesitant to offend sensibilities, especially his wife's, but he rejected the
notion that the view of life he offered lacked the dignity attached to it

by religion. The concluding words of his book show how much awe he felt:

> There is grandeur in this view of life, with its several powers, having been originally breathed into a few forms or into one; and that, whilst this planet has gone cycling according to the fixed law of gravity, from so simple a beginning endless forms most beautiful and most wonderful have been, and are being, evolved.

Nevertheless, he long delayed the publication of his manuscript until finally persuaded to guard the priority of his revolutionary new insight when he became aware that the naturalist Alfred Russel Wallace had similar ideas and was about to publish them. *On the Origin of Species by Means of Natural Selection* was finally published in

Figure 2.2.   A cartoon about the theory of evolution published in *Harper's Weekly*, August 19, 1871.

1859, followed twelve years later by *The Descent of Man, and Selection in Relation to Sex.* Both books instantly became famous, and the general reaction was tumultuous, filled with antagonism. Most of the objections were based on religious grounds or on what is today called 'intelligent design' and therefore impossible to refute by rational argument — though some, such as the botanist Joseph Hooker and the biologist and paleontologist Thomas Huxley, as well as others, tried — but there were also some serious scientific difficulties.

It was clear that evolution driven by natural selection was an exceedingly slow process that would require a vast number of generations and thus many millions of years to lead to the evidence of extinct species in the fossil record and to the variety of life today, including humans. Was the Earth old enough for this? If one believed in a literal interpretation of the Bible, the answer was, of course, no: Genesis said it was no older than about 6000 years. But there also seemed to be well–grounded scientific arguments that the Earth was too young to allow for the slow pace of natural selection to drive the evolution of species all the way to *Homo Sapiens.* Lord Kelvin, one of the most highly esteemed contemporary physicists, had estimated the age of the Earth by assuming it started in the form of hot molten rock and then slowly cooled. His calculation came out with the result that the Earth was between 20 and 400 million years old, with a most likely age of about 100 million years, which was not nearly sufficient for Darwinian evolution.

Fortunately for Darwin's idea, it turned out that the discovery of radioactivity at the end of the nineteenth century fundamentally altered our knowledge of the heating processes involved. Radioactivity in the Earth's interior provides an internal source of heat, and if that as well as the heating by the sun's rays are taken into account, the cooling process of the Earth is drastically slowed down and its age is now calculated to be approximately 4.54 billion years, which is plenty of time for Darwin's theory. The theory has also been bolstered in the meantime by the discoveries of the laws of heredity by Gregor Mendel as well as of the biochemical mechanisms of genetics by Francis Crick and James Watson. The combination of Darwin's theory with our present knowledge of genetics is known as the Modern Synthesis of evolution and it now forms the foundation of the science of biology.

Charles Darwin died in 1882 and was given the rare honor of a state funeral at Westminster Abbey, buried not far from Isaac Newton.

Gregor Mendel

Born in 1822 as the son of poor peasants in the village of Heinzendorf in Austrian Silesia (now part of the Czech Republic), Johann Mendel showed himself at a young age to be so intelligent that his mother hopefully envisioned a better future for him than the miserable life of his parents. This, however, required that he be educated, and they had little money to provide him with the necessary schooling. They did scrounge up enough, though, for him to attend elementary school and even, with great sacrifices of comfort, high school in nearby Troppau, where he did extremely well. He might indeed be able to escape the life of a peasant and perhaps become a teacher. But for that he would have to go and take courses at a place like the Philosophical Institute at Olmütz, which would cost money his parents did not have. With the help of his sister and some money earned by tutoring, he managed to do this, but that was as far as he could go. Without any well connected friends or relations he would never be able to find a teaching position.

Young Mendel loved science, especially physics and mathematics, and would like nothing better than to devote the rest of his life to it. One of his teachers, who recognized that Johann had an excellent mind, finally came up with a suggestion: how would he like to join the Augustinian monastery at Altbrünn and become a monk? He might be able to live there and study as well as teach at a nearby school. A fantastic idea; his mother had already suggested long ago that she would love for him to become a priest, and now it might become reality.

This is how Johann Mendel entered the Augustinian order of St. Thomas, adopting the name Gregor Johann Mendel, as it was customary for monks to take a new name. At the age of twenty-five he became an ordained priest, and a year later Pater Gregor was temporarily appointed to teach mathematics as well as Latin, Greek, and German literature at the High School at Znaim in Moravia. In order to enable him to make his teaching position permanent, the monastery arranged for him to enroll at the University of Vienna for a couple of years, where he enthusiastically took courses in various sciences, including physics with the famous Professor Doppler (discoverer of what became known as the Doppler effect in acoustics). Even though he twice failed his examination for a teaching certificate — he stubbornly refused to agree with the arrogant examiners on some questions relating to heredity in the breeding of flowers — Mendel was appointed to a 'temporary' teaching position at the Modern School in Brünn (now Brno), which he nevertheless retained for many years.

From then on Pater Mendel lived at the local Augustinian monastery near Brünn and taught at the Modern School. As an enthusiastic gardener he adopted the hobby of breeding hybrid flowers, which he could very conveniently pursue in the beautiful gardens of the monastery. This soon turned into more than a hobby: he began asking himself very specific questions to which he could find no answers in the literature. Why were the results of specific hybridization experiments so stable, always coming out the same? But when he allowed the seemingly equal hybrids to fertilize one another or to self-fertilize, the results in the next generation were not stable. Apparently the hybrids, though they looked the same, were not equal after all. Indeed, he was told, it has been known to breeders for a long time that hybrids, both among plants like peas and animals like pigeons and rabbits, did not breed true. His view of the natural world being closer to that of a chemist or physicist, Mendel did not empathize with botanists' apparently satisfied acceptance that living things were simply not predictable. He made it his task to conduct systematic experiments to get some answers to these questions. What he was planning to do was to actually keep track and count the results of hybridizations from one generation to another. He realized that in order to discover the underlying laws of heredity, if there were any, he would have to count the results of thousands of carefully controlled hybridizations from one generation to another.

The plants he decided to work with were ordinary garden peas, a species that is naturally self-fertilizing, relatively easy to cross-pollinate by hand, and whose hybrids are perfectly fertile. He had to make sure he started with seeds that bred true and were not contaminated by hybrids, that seeds in bags labeled 'yellow peas' really did produce only yellow peas, and not only for one generation; similarly for those labeled 'green peas' or 'dwarf plants,' etc. To be certain, he first tested all the seeds he used, even though they came from the most reliable sources he could find, for a couple of seasons to make sure they bred true. In each set of experiments he meant to pay attention only to one of seven pairs of traits (even though every individual plant would of course exhibit more than one of these trait pairs): 1. shape of the seeds (smooth or wrinkled); 2. color of the seeds (yellow or green); 3. color of the seed coats (white or grey); 4. shape of the ripe pods (inflated or constricted); 5. color of the unripe pods (green or yellow); 6. position of the flowers (distributed or bunched at the top); and 7. length of the plant's stem (tall or dwarf).

Then he set to work in his extensive garden, sectioned off for the various trait tests. What he had to do was simple enough but extremely tedious: each flower had to be pollinated by hand. Before any pollen had been shed, the stamens of each young flower to be cross pollinated had to be clipped off, and the pollen taken with a fine brush from the flower of another pea plant had to be transferred to it. Finally, in order to prevent any other uncontrolled pollination by insects or the wind, a small calico bag had to be tied around each artificially pollinated flower. The first series of experiments consisted of almost 300 cross-fertilizations on 70 plants, and the maturation time was about eighty days. He now had the seeds of his first-generation hybrids, and these had to be planted the following spring. Only after they had fully grown would he be able to see what the hybrids actually looked like.

The results of this first generation of hybridization, he discovered, were quite surprising. The parental characters were not randomly distributed among their off-spring; instead, in every instance, one member of each pair of parental traits overwhelmed the other: all the seeds from the tall-dwarf hybrids grew into tall plants; all the inflated-constricted hybrid seeds grew into plants bearing only inflated pods, etc. One of the two traits was always dominant and the other recessive, as he called them.

His results, then, for the first-generation hybrids were: 1. smooth seeds dominant over wrinkled; 2. yellow seeds dominant over green; 3. gray seed coats dominant over white; 4. inflated pods dominant over constricted; 5. green color of the unripe pods dominant over yellow; 6. distributed blower bunches dominant over bunched at the top; and 7. tall plant stems dominant over dwarf stems. He later discovered that it was not true for all kinds of plants that one trait always dominated over the other; sometimes hybrids had a mixed appearance: a hybrid of red flowers with white flowers might be pink. Nevertheless, Mendel's finding of dominant and recessive traits in heredity was a monumental discovery.

The next step was to allow the first-generation hybrids to self-fertilize, by simply tying calico bags around each flower before they were mature, so that no external pollination could occur. He then made an exact tabulation, for each of his seven character pairs, of what the characteristics of the second generation were. Some of the off-spring of a smooth-wrinkled hybrid (which themselves were all smooth) were smooth, while others were wrinkled — even though the recessive trait never appeared in the first generation hybrids, it reappeared in the second generation. This was the well-known fact that hybrids did not always breed true. He realized that he had to keep careful track of the numbers in order to be able to draw scientific conclusions; indeed the crux, as he saw it, was the ratio of one trait to the other exhibited in the second generation hybrids. And similarly for all seven of his trait pairs. His botanist friends were quite puzzled by all this and did not understand what Mendel was up to.

Mendel's first results already showed what he regarded as significant: the ratio of the number of plants having dominant traits to the number with recessive ones in the second generation hybrids came out to be 3.01 to 1. If his further experiments confirmed that this numerical ratio was an experimental approximation of what was really 3 to 1, as he suspected, then he thought he had discovered something important. And indeed, when he counted the results of all the thousands of offspring of self-fertilized first-generation hybrids of his seven trait-pairs, he found dominant to recessive ratios that varied between 2.84:1 and 3.10:1; and there was no discernable difference in these ratios between any of the trait-pairs. Rather than being satisfied, as most botanists at the time would have been, that this ratio fluctuated in an uncontrolled manner — life was unpredictable — he concluded that in the second generation, the dominant trait showed itself 75% of the time and the recessive trait 25% of the time.

What could be the underlying reason for this fixed ratio? After much thought, Mendel developed a theory: the source of the visible traits had to be in specific sex cells in each pollen and each egg. As he saw it, these sex cells of hybrids contain both the factors for the dominant and the recessive trait, but during fertilization these factors must separate, and only one of the factors from the pollen combines with one of the factors from the egg, without preference as to which of them. (Exactly what the 'factors' actually consisted of, he had of course no idea.) There are then four ways for the pairs of factors from pollen and egg to form the new cells: dominant-dominant, dominant-recessive, recessive-dominant, and recessive-recessive, and

Figure 2.3. Gregor Mendel.

the choice would be random. The first three lead to a plant with the dominant trait, and only the last to a plant with the recessive trait: exactly the 3:1 ratio he had observed. Furthermore, the first and the fourth combination of factors would lead to plants breeding true — the first with the dominant trait, the fourth with the recessive one — whereas the second and third to plants that were like the first generation hybrids. In other words, among the second-generation hybrids, all the plants showing the recessive trait would always breed true, whereas those exhibiting the dominant trait would not. All these conclusions he subsequently verified by more experimentation.

This brilliant and totally unexpected insight is now called Mendel's Law of Segregation. It took another fifteen years before chromosomes were discovered and suggested as the physical carriers of the hereditary trait-'factors' found by Mendel. Another twenty-five years passed before it was established that indeed new sex cells are formed from the splitting of paired parental chromosomes as Mendel had envisioned abstractly with no benefit of knowledge of the cellular mechanisms.

But all that was in the future. Mendel was convinced he had found a fundamental law of heredity, and he kept on experimenting with his peas for years to verify the predictions implied by his ideas for further and further generations of hybrids and their off-spring. He then discovered another important fact: the heredities of different trait pairs in plants did not interfere with each other; they were quite independent. This became known as the principle of independent assortment, or Mendel's second law.

As time went on, Mendel learned about Darwin's recently published work, *On The Origin of Species by Natural Selection,* and discussed it with others. The Lamarckian explanations, which Darwin sometimes relied upon, he concluded were of no value in light of his own results. He was convinced that what he had found played a fundamental role in understanding how the hereditary changes Darwin postulated came about and closed some gaps in Darwin's theory.

After eight years of experimenting with garden peas, Mendel finally wrote up his results and presented a paper in 1865 to two meetings of the Brünn Society for the Study of Natural Science and published it in the Society's Proceedings. His friends congratulated him, but no one understood the significance of what he had done. While waiting for some reaction to his publication, a copy of which he had sent to a famous professor he knew to be an expert in the subject, Mendel continued with hybridization experiments using different kinds of plants. To his great disappointment, however, the general response was silence and even several years of exchanges of letters with the famous man elicited nothing positive or encouraging. Though he continued to believe that he had discovered very important principles of heredity, Mendel lost faith that this would ever be recognized during his lifetime; sadly for him, this turned out to be correct.

When the prelate of the monastery of St. Thomas died, Mendel allowed himself to be persuaded to stand in the election for a successor. So in 1868, Pater Mendel, the son of poor peasants, became Prelate Mendel, abbot of a very wealthy Augustinian monastery. However, he continued to pursue scientific studies of plants in the little free time left for him by the heavy burden of administrative work. He died in 1884, widely honored and loved in his community, but almost no one, including Charles Darwin, knew of his path-breaking contribution to the science of genetics.

It was not until the year 1900 that the Dutch botanist Hugo de Vries, who had performed experiments on hybridization and published his results, found to his surprise a reference to Mendel's monograph published 34 years earlier in a well-researched bibliography, which already contained all the results he had obtained. And shortly thereafter a German botanist, Carl Erich Correns, who had worked on pea hybrids for years, published a paper entitled

Figure 2.4. The memorial to Gregor Mendel erected in Brno in 1910.

Gregor Mendel's Rules Concerning the Behavior of Racial Hybrids. The same year, an Austrian botanist by the name of Erich Tschermak also published a paper on pea hybrids, paying enthusiastic tribute to the Augustinian monk. The work of Gregor Mendel was at last recognized sixteen years after his death. Finally, in 1910, a memorial was erected in his honor in the city of Brno (see Fig. 2.4).

Louis Pasteur

Born in the Jura region of France, Louis Pasteur grew into his teens at the time when Darwin returned from his voyage to Patagonia and the Galapagos Islands. After being educated at the Ecole Normale Superieure in Paris — the best institute of higher education in France originally founded by Napoleon — he had a burning ambition to make a lasting contribution to science in the field regarded at the time as the most fundamental, which was chemistry. But he began his career by experimenting in crystallography, frequently spending much of his own small salary as a teacher on the needed equipment and material. He first made his name by performing detailed experiments that established the varying light-transmission of crystals, especially their effects on the polarization of light, a property that had been discovered not long before. Every pure monochromatic light beam can be decomposed into a vertical and a horizontal polarization component. The instrument that measures the direction of the polarization of incoming light is called a polarimeter, but nowadays the familiar Polaroid sunglasses serve too; they transmit only one of the polarization components. If you rotate them, the polarization of the transmitted light is rotated as well. When polarized light shines through crystals, some of them, Pasteur found, always automatically rotate its polarization to the right, others to the left. Such crystals therefore have an inherent handedness: some are right-handed, others left-handed.

Carrying this background with him when his interests changed to chemistry, he discovered to everyone's astonishment that sometimes two substances with the identical chemical composition nevertheless formed crystals of the opposite handedness. There had to be more to a chemical compound than is expressed in its chemical formula, which simply tells which elements are contained in it and how many atoms of each element the molecule of the compound consists of. He thus founded what became known as stereo-chemistry, which pays attention not only to the number of various elementary atoms in a molecule, but also to their spatial arrangements. These special configurations were not always symmetrical: there were left-handed molecules and right-handed molecules. Sometimes two molecules made up of the same atoms differed in their asymmetrical arrangement; for example, there is right-handed sugar and left-handed sugar. What is more, his exacting experiments convinced him that the compounds whose molecules possessed such handedness properties were always somehow connected with life!

From then on Pasteur changed the direction of his scientific work towards biology, using both the chemist's test tube and the biologist's microscope for detailed experimentation and observation in the new science of biochemistry he had created. The concrete impetus for this change was his attempt to understand fermentation, a chemical process that produced wine and beer but also sometimes caused very damaging spoilage. What was the nature of the action of yeast during fermentation and in the leavening of bread? He discovered what was involved here was not merely straightforward chemistry; the presence of tiny microscopic animals was crucial. Some of these animalcules, subsequently called germs, bacteria or microbes, which had first been seen in the seventeenth century under the microscope by Antony van Leeuwenhoek, needed oxygen to live (as was expected of all life) but others needed no oxygen, he discovered; the former he called aerobes, the latter anaerobes, though the borderline between the two kinds was not always sharply defined. Pasteur's future work would require not only a microscope in addition to the test tube but also an incubator for the breeding and multiplication of bacteria.

The first tangible result of Pasteur's discovery of the presence of microbes in fermentation and their strange properties was that he was able to cure a fermentation-related disease of wine that had caused great losses to many vineyards. The cure he found consisted of a very carefully controlled and regulated heating process to a temperature just below the boiling point, which killed the bacteria without damaging the wine. This procedure, eventually called pasteurization, turned out to be extremely useful in preventing the spoilage caused by the presence of microbes of many other perishable fluids such as milk, and it is of course employed to this day.

It was inevitable that Pasteur would become embroiled in the heated controversy raging at the time about spontaneous generation. It had long been thought that living creatures could be generated from inorganic matter: worms would grow by themselves in apples,

Figure 2.5. Louis Pasteur.

maggots in decaying flesh, flies in fermenting meat, even mice could be born in dirty rags, all without the prior presence of eggs or other forms of life. Pasteur devoted large amounts of time and effort, and many ingenious experiments, to prove this belief to be wrong: life could not be generated out of nothing. But although he managed to disprove all alleged instances of such spontaneous generation that he encountered, he never could silence this false idea once and for all. It eventually simply died from too much counter-evidence.

Becoming well known as the savior of many French vineyards by his heating procedure, the young Pasteur was asked by the government to help with a serious disease that was afflicting France's sericulture. The large French silk industry in the Midi depended for its existence on the extensive growing of silkworms, caterpillars which produce the remarkably strong thread woven into silk fabric just as spiders produce the thread from which they spin their web. An unknown disease suddenly began to kill large numbers of these worms. Armed with his microscope, Pasteur discovered that the cause of these deaths was a tiny parasite transmitted by the silkworms' excrement left on the mulberry leaves they fed on. The cure of the disease he suggested was drastic, and nothing less would do: the killing of all existing silkworms and starting from scratch with specially selected healthy worms. It worked, and it saved the entire French silkworm industry. At the same time, Pasteur had his first experience with the contagious transmission of a disease by microscopic germs.

This newly acquired knowledge was soon applied more generally. If invisible germs were responsible for transmitting disease, great care should be taken by physicians in contact with open sores and wounds, as well as by midwifes, as the Hungarian physician Ignaz Semmelweis in Vienna had strenuously but ineffectively urged on the basis of his own recognition of the cause of infections by contagion. The Scot Joseph Lister was the first physician to apply Pasteur's discoveries to medicine, recognizing gangrene as a pathological form of fermentation — putrefaction — in humans, caused by bacteria transmitted either through the air or by direct contact with an external agent. Both Pasteur and Lister started a concerted campaign that lasted for about ten years to teach physicians the basic rules of antiseptics, the spraying of carbolic acid (later modified to other antiseptics), the wearing of gloves and protective clothing, the sterilization of all operating instruments in an autoclave, and above all, the thorough washing of hands after each contact with a patient. It was not an easy sale to the medical establishment, which resisted the bothersome new precautions they regarded as unnecessary. However, the results were clearly apparent: morbidity due of infections after childbirth, compound fractures, amputations, and other surgery fell drastically.

Pasteur's next medical adventure in the struggle against contagion was the use of vaccination in the fight against anthrax, the disease that often devastated large herds of sheep, cattle, as well as horses, that could also be transmitted to humans with deadly effect. Robert Koch had already demonstrated that anthrax was caused by micro-organisms in the blood. Moreover, these micro-organisms could survive for years in the form of spores in the soil where the cadavers of anthrax victims had been buried, thus infecting animals grazing there later. Pasteur found that the disease could also be transmitted by means of anaerobic germs in the body rather than the blood, and that chickens infected with these germs did not succumb to it. His very careful experimentation established that it was the high temperature of the chickens' blood that killed the anaerobic germs, and at the same time that the chickens so infected were immune to the disease even when plunged into a cold bath that lowered their blood temperature. This was the crucial observation leading to his development of a vaccine.

The idea of conferring immunity to certain diseases by vaccination was not new at the time of Pasteur. Rudimentary forms of it had been practiced for ages against the dreaded smallpox, for example, by infecting people mildly by means of the puss from smallpox pustules, taken at a certain stage of development of the disease. Edward Jenner in England had used cowpox germs from a cow's udder to infect humans with this relatively mild disease, after which, he found, they were immune to smallpox. But he had a very hard time persuading his medical colleagues to use this method as a prophylaxis, especially since it did not always work.

In his fight against anthrax Pasteur concluded from his experience with chickens that for the production of a vaccine it was crucial to have an attenuated parasite, and this could be achieved, as he found after extensive experimentation, by carefully raising the temperature of its growing culture to between exactly 42° and 43^oC for eight days while infusing it with oxygen. The resulting microbes had lost their virulence and were harmless to rabbits, sheep, and cows, but their injection made these animals immune to anthrax. He had invented a simple but reliable method of producing an effective vaccine. (His explanation of how these microbes lost their virulence was that it had to be a Darwinian change of species, or possibly an adaptation to its environment by learning in the form taught by Lamarck.)

Pasteur's final dramatic conquest was the defeat of a terrifying human disease, several cases of which he had witnessed as a child and could never forget: rabies. Even though under his microscope he could not see the microbe transmitting it — he called the tiny germ a virus — he assumed it had to exist in the brains of hydrophobic dogs or other rabid animals whose bite caused the disease in humans, which after several weeks of agonizing incubation almost invariably leads to their death. The great challenge in this case was to find not only a prophylactic vaccine for dogs, but a cure once a person had been infected.

Relying on his previous experience with an anthrax vaccine, Pasteur found a way to weaken the virulence of the rabies virus sufficiently to act as a vaccine preventing dogs from becoming rabid.

Figure 2.6. Sheep being vaccinated against anthrax at Pouilly-le-Fort in 1881.

But this was not sufficient for curing a human after a bite by a rabid animal. He had the brilliant idea that what he needed to produce was an attenuated virus with a short incubation period. This could then be injected after an infection by a bite — such an infection had a relatively long incubation period — and stimulate a relatively quick response, thus acting like a vaccine before the original infection had incubated and become effective. It took a long period of difficult and painful animal experimentation, which we need not describe in detail. Suffice it to say that he finally had to have this risky treatment administered by a physician in a human being to find out if it worked. (Because Pasteur was not a medical doctor, he was not allowed to perform such injections himself.)

The crucial test case turned out to be a nine-year old boy by the name of Joseph Meister, who was brought to Pasteur by his mother on July 6, 1885, shortly after being mauled and severely bitten in several places of his body by a rabid dog. There was no question but that he would almost certainly die without Pasteur's experimental procedure. Under his careful supervision, several of the new fast-acting vaccines were injected successively over a period of ten days with increasingly virulent viruses. By July 27, Joseph was sent home, feeling cheerful and well, and Pasteur felt fully reassured: the boy's life had been saved and rabies had been defeated, though he would have to wait some time to be sure. Pasteur stayed in touch with Joseph for years, even helping with his education and finding a job for him as he grew up, a living testament to the effectiveness of Pasteur's rabies treatment.

Another test of the new procedure soon followed: a shepherd, trying to prevent a rabid dog from attacking his sheep, had been bitten himself. After he was brought to Pasteur's laboratory, the same protocol that had saved the life of Joseph Meister was administered, and again it succeeded. Since this case did not occur during the sleepy summer holidays as had Joseph's, the saving of the heroic shepherd became instantly famous. Pasteur was celebrated all over France and many new cases were brought to his laboratory for his miraculous life-saving treatment. His fame even spread to Russia, where nineteen people who had been badly bitten by a rabid wolf were sent to Pasteur in Paris. He managed to save all but three, whose arrivals had been delayed too long. Confidence in the safety of his rabies vaccine was so great that his assistants had themselves injected with it as human guinea pigs. (Pasteur volunteered himself but they refused to try it on the master.)

Laboratories for rabies treatment by Pasteur's method were soon instituted not only in St. Petersburg but also in London, Vienna, Jena, and Warsaw. In Paris, the Institut Pasteur was set up, funded by private subscription with contributors from many countries. It would eventually become one of the greatest biological research institutions in the world, as was appropriate for the commemoration of a great scientist who had made many other important contributions to biology not described here. Louis Pasteur died in 1895 and was accorded a state funeral at Notre Dame de Paris. His burial crypt is at the Institut Pasteur.

Michael Faraday

Michael Faraday grew up with almost no elementary education, a son of a blacksmith in Newington, Surrey. At the age of thirteen in 1804, as an errand boy and then apprentice bookbinder in London learning a trade, he had his first contact with books, which he devoured. His intellectual fires were especially kindled and directed toward science when he learned about electricity by reading the Encyclopaedia Britannica. More reading, attending lectures, and helping to organize a discussion group in order to exchange ideas with other young men orally as well as in letters, added to his knowledge and understanding not only of science, but also in his general education and the use of correct English, in which he felt himself to be deficient. His self-confidence soon improved to the point of delivering his first lecture at the City Philosophical Society at the age of nineteen. His subject was electricity, theories about which were roiled by controversy, and he did not beat about the bush but took a firm stand on the side that held it consisted of two fluids, one of positive and the other of negative electricity.

This was a time when the great chemist Humphry Davy regularly lectured at the Royal Institution to large popular audiences, which sometimes included young Faraday, persuasively arguing that chemistry was the most fundamental of the sciences. Many of the latest experiments were of an electro-chemical nature and hearing Davy tell about them induced Faraday to redirect his primary interest to chemistry. At the age of twenty-one, he finally had an enormous stroke of luck, which came in the form of an accident in Davy's laboratory that temporarily blinded the chemist. Recommended by a friend, Faraday was hired for a few days as Davy's amanuensis and, after impressing Davy with the notes he had taken at his talks, he was kept on as an assistant in his laboratory at the Royal Institution, of which Davy had become the Director. Young Faraday now turned into an enthusiastic chemist, undeterred by some of the explosive dangers of Davy's experiments.

Shortly after being knighted and newly married, Sir Humphry Davy planned to embark on a grand tour of the European continent with his wife, and he decided to take his brilliant young assistant along with them. For Faraday the voyage, with Davy as his tutor, became the equivalent of a university education that he never had. Not only were his eyes opened to entirely new social experiences, and he acquired a working knowledge of French and Italian, but since Davy had brought along a small chemical laboratory chest, they immediately performed experiments stimulated by the latest scientific news. Perhaps the most valuable experience of travelling in the company of the most famous chemist in Europe was that Faraday made the acquaintance of many other prominent scientists, such as Volta, Humboldt, Cuvier, Gay-Lussac, and Count Rumford.

Figure 2.7. Faraday as a young man.

Upon his return to London, Faraday was offered a new job at the Royal Institution: he became Assistant and Superintendent of the Apparatus of the Laboratory and Mineralogical Collection, and he moved into apartments in the building of the Institution that his position entitled him to. Within a few years he was not only working on his own instead of depending on Davy, but also rapidly emerged as the foremost analytical chemist in Britain, especially on alloys and clays, with pioneering research on steel alloys. Branching off into organic chemistry, and disdaining the notion of special vital forces popularly seen at the time as operating within compounds containing carbon, he established the chemical formulas for two new carbon compounds — tetrachloroethene and hexachloroethene — as well as a new compound of carbon, hydrogen, and iodine. He also established the laws governing electrolysis and discovered benzine.

At the same time, however, Faraday began to be intensely interested in physics. (He never regarded himself as either a chemist or a physicist but as a natural philosopher.) After coming across a paper by the Danish physicist Hans Christian Ørsted, which showed that an electric current flowing in a wire deflected the needle of a nearby compass — thus demonstrating that electric currents produced magnetism — he began experimenting on the interaction between currents and magnets in his own way. Forming an electric circuit by immersing the end of a wire in a vat of mercury for mobility and putting a bar magnet in the vat (see Fig. 2.8), he found that when the wire was held fixed, the end of the bar magnet rotated about the current-carrying wire, and when the magnet was held stationary, the wire would rotate about it. He had found a remarkable phenomenon: electromagnetic rotation. What was so unusual about this, as well as about Ørsted's discovery, was not that electric currents produced forces — the French scientist André Marie Ampère had already demonstrated that current-carrying parallel wires exerted forces on one another — but that the electromagnetic forces were not central forces like Newton's gravity, pulling in the direction of its

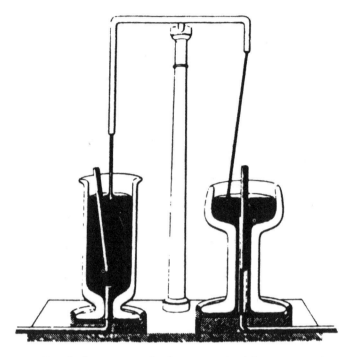

Figure 2.8. Faraday's apparatus for demonstrating 'electromagnetic rotation.'

source. Ampère's force was at right angles to the currents and Ørsted's as well as Faraday's force between currents and magnets made objects rotate about one another.

Faraday's most important discovery, however, went beyond the forces exerted between electric currents and magnets. He found that not only did currents in wires make nearby magnets move, the inverse effect also occurred: a magnet moving near a wire, or any magnetism of time-varying strength, produced an electric current in the wire. Called electromagnetic induction, this effect — unbeknownst to Faraday, it had actually been discovered shortly before by the American physicist Joseph Henry but Faraday was the first to publish it and the law governing it was named after him — would turn out to have many important applications. In fact, most of the electrical power-devices driving nineteenth- and twentieth-century industry, such as transformers, dynamos, electric motors, and electromagnets, were based on discoveries by Michael Faraday.

His contributions to physics were not limited to experimental discoveries. Beginning to think seriously about the nature of electromagnetic forces and how to describe them, he did not want to make use of the intellectually repellent idea of action at a distance that Newton had introduced for the force of gravitation. (Nevertheless, Faraday tried unsuccessfully to find a connection between the electromagnetic forces and the force of gravitation.) In his very visual imagination he saw the forces exerted between electric charges, between magnets, or between currents and magnets as transmitted by 'lines of force' like rubber bands, and he saw all space filled with them. In the case of magnetic forces, they could actually be made visible by iron filings surrounding magnets (see Fig. 2.9). The totality of these space-filling lines of force he called an electromagnetic field. Unfortunately, Faraday's mathematical ability was not sufficient to capture this notion, one of the most fertile ideas in the history of physics, in the form of equations; it took James Clerk Maxwell to accomplish that some years later. In fact, Maxwell was indebted to Faraday, which he clearly acknowledged, not only for the invention of the electromagnetic field, but also for the discovery that the polarization of light can be rotated by a magnet. This indicated for the first time that light is an electromagnetic phenomenon, which Faraday explicitly speculated upon and which Maxwell's equations incorporated: light is an electromagnetic wave. To this day, generalizations of Faraday's field notion dominate fundamental physics.

In the meantime, Faraday the natural philosopher also became an extremely successful public lecturer. He inaugurated the annual Christmas Lectures for young children as well as the Evening Discourses for members of the Royal Institution and their guests. His popularity eventually surpassed even that of Sir Humphry Davy in his heyday. Unlike the latter, however, he disdained all public honors.

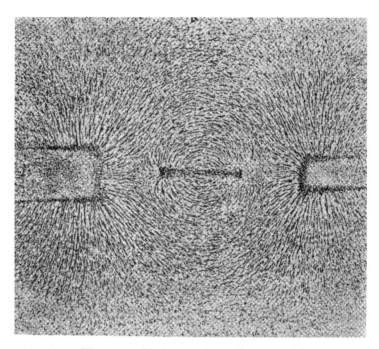

Figure 2.9. Iron filings sprinkled on a sheet of paper, with a magnet placed beneath.

The strain of his intense work load was beginning to show by the time he turned fifty, when he had a breakdown. In 1841 he took off and spent eight months with his wife and brother walking in the Swiss mountains, returning refreshed and productive. He discovered that magnetic fields influenced not only such substances as iron but also many others, such as glass, which were subsequently called diamagnetic, because their reaction to magnetism differed from those that were dubbed paramagnetic. The latter were like the so-called ferromagnetic materials iron, cobalt, and nickel but did not retain their magnetism like the ferromagnets. His discovery was the beginning of the discipline of magnetochemistry. As time went on, however, his memory began to fail him and his mental powers deteriorated (possibly the long-delayed result of some accidental poisoning during his early chemical experiments). At the age of seventy-one he resigned his position at the Royal Institution and Queen Victoria provided him with an apartment at Hampton Court, where he retired. Michael Faraday, the man whom many scientists consider to have been the greatest experimenter of the nineteenth century, died in 1867, at the age of seventy-six. According to his wishes he was buried at Highgate cemetery in a very private ceremony.

Max Planck

More than anyone else, Max Planck personified the transition from what we now call 'classical physics' to modern physics. Born in 1858 in Kiel, a city in northern Germany, he descended from a long line of theologians and a father who was a professor of law at the local university. After studying physics — though being advised by one of his prominent teachers that physics was a field in which nothing new could be expected as everything had already been discovered — at the universities of Munich and Berlin, he was appointed as professor of theoretical physics first in Kiel and soon thereafter in Berlin.

Planck's specialty was thermodynamics, the science of heat, and in the late 1890s this still developing field had to confront an unsolved puzzle: how to understand the way the frequency distribution of the radiation emitted by a 'black body' (a totally absorbing, non-reflecting object) depended on its temperature — think of a glowing hot piece of iron and how its color changes as it is further heated. (The object could also be a cavity with a tiny hole in it to allow some of the radiation to escape.) The physicist Wilhelm Wien had derived what became known as Wien's Law from classical thermodynamics, but this law failed to agree with experimental results.

Even though Planck was not yet quite convinced by the new-fangled view of thermodynamics based on the statistical mechanics of atoms that make up all matter (recently proposed by Ludwig Boltzmann — see the next chapter), he decided to make use of it. What is more, he made the ad hoc assumption that the energy of the radiation emitted by the atoms of the black body (or the walls of the cavity) always came in the form of small packets, each of which had an energy E proportional to its frequency ν, E $=$ hν, where h is a universal constant. It turned out that the resulting replacement of Wien's Law by Planck's Law agreed with experiments, and the reputation of its discoverer was forever firmly established. Max Planck was now the most famous physicist in Germany.

Five years later, the young patent clerk Albert Einstein in Basel published three papers, two of which shook physics to its foundation: one contained the special theory of relativity while the other initiated the quantum theory of radiation by universalizing Planck's quite restricted assumption and postulating that all electromagnetic radiation came in the form of quanta — later called photons — whose energy E was proportional to its frequency ν by Planck's equation E $=$ hν. Very few physicists paid attention to or understood Einstein's path breaking papers, but Max Planck recognized the genius of their author and invited him to come to Berlin as a professor.

The thoroughly conservative Max Planck had inadvertently triggered a revolution. Well aware that his *ad hoc* procedure of deriving the successful black-body radiation law was unconventional, he nevertheless had no intention of changing any basic assumptions of accepted physics. It remained for the young iconoclasts Einstein, Niels Bohr, and Werner Heisenberg to launch the revolutionary new quantum theory, now the most basic explanatory tool of all submicroscopic physics, with Planck's constant as its cornerstone. Uncomfortable as Planck was with these new ideas, when they led to well verified results, as indeed they did, he accepted them without quarrel or active resistance.

As time went on, Max Planck became and remained the great man of German science. Conservative nationalist that he was, during the First World War he signed — to his deep later regret — the infamous Manifesto of 93 prominent intellectuals, which denied any German war guilt and misconduct during the invasion of Belgium (Einstein refused to sign). His stature was such that even the Hitler regime did

Figure 2.10. Max Planck.

not dare touch him when he refused to join many of his colleagues in becoming a Nazi or when he quietly helped some Jewish scientists (though he never openly opposed the regime). He lost his eldest son as a soldier during the First World War, and his second son was executed by the Nazis, accused of being involved in the unsuccessful 1944 attempt on Hitler's life.

At the war's end, the little that remained of the tattered honor of German science after twelve years of barbarism was personified in Max Planck, and he died in 1947 at age of 89. All the German scientific institutes that had long been named after Kaiser Wilhelm were renamed after Max Planck, and today his name, attached to a fundamental length, a fundamental energy, and a fundamental mass, is certain to be referred to in almost every publication of research in particle physics.

Enrico Fermi

Born in Rome in 1901, and showing an early aptitude for mathematics and physics, Fermi was educated at the Scuola Normale Superiore in Pisa and earned his doctor's degree in physics at the University of Pisa. After some post-doctoral research in Göttingen and Leyden, he returned to Italy to take the position of Lecturer in Mathematical Physics and Mechanics at the University of Florence. While there, concentrating mostly on the new fields of quantum mechanics, atomic physics, and general relativity, he discovered the statistical laws governing particles that obey the Pauli exclusion principle (see the next chapter); as these laws were simultaneously discovered by the British physicist Paul Dirac, they were named Fermi-Dirac statistics. The particles obeying them are now called *fermions*. The next year, 1927, Fermi moved back to his birthplace to become Professor of Theoretical Physics at the University of Rome.

After James Chadwick discovered the neutron in 1932, it was found that beta-radioactivity appeared to violate the law of conservation of energy, a calamity that Wolfgang Pauli averted by suggesting the emission, in the same process, of a hitherto unknown almost weightless neutral particle, which Fermi named *neutrino* — the little neutral one. The name stuck. Since the neutron itself turned out to be subject to beta-radioactive decay, Fermi turned his attention to the problem of explaining the nature of this fundamental process. The theory of beta decay that he invented was highly regarded for a number of years but it was eventually superceded by the electroweak theory because some calculations with Fermi's theory led to nonsensical infinities.

He then turned his full attention to nuclear physics. Stimulated by Irène Curie and Frederic Joliot's discovery of artificial radioactivity, he began to do actual experiments himself and found that bombarding elements with neutrons led in almost every instance to nuclear transformation, sometimes, it turned out, producing radioactive isotopes of elements, and sometimes elements heavier than any in

Figure 2.11.   A photograph of Enrico Fermi.

the periodic table, so-called transuranic elements. What's more, he discovered the important fact that slow neutrons were much more efficacious for this purpose than fast ones. Chafing under Mussolini's dictatorship, and particularly under the newly introduced antisemitic laws — his wife was Jewish — he took advantage of his trip to Stockholm to accept his Nobel Prize in 1938 and emigrated with his wife to the United States. (On the way, to hide the valuable gold medal that came with the prize, he dissolved it in a jar of acid, later precipitating the gold, though of course not in the form of the medal.) Soon after his arrival, he was appointed Professor of Physics at Columbia University in New York.

When the German chemists Otto Hahn and Fritz Strassmann, trying to produce transuranic elements by slow-neutron bombardments of uranium as Fermi had recommended, accidentally discovered what turned out to be nuclear fission in 1939, Fermi immediately realized the possibility of a chain reaction produced by the emission of neutrons in each such event and the enormous potential of this process. He was put in charge of attempting to construct the first nuclear reactor — at the time it was called a pile — by producing just such a chain reaction but controlling it before it could lead to an explosion. The experiment he set up in a squash court at the University of Chicago in 1942 succeeded. (It was exactly this feat that the German physicist Werner Heisenberg, the author of the famous uncertainty principle, attempted for Hitler's Germany but fortunately never accomplished.) The rest of the war he spent as a leader under Robert Oppenheimer of the Manhattan Project at Los Alamos, New Mexico, successfully developing the generation of nuclear energy and the atomic bomb. After becoming an American citizen, he accepted a position as a Professor at the Institute for Nuclear Studies of the University of Chicago, where he was a very popular teacher pursuing research mostly in high-energy physics and cosmic rays.

Enrico Fermi died in 1954. He had been that rare kind of physicist who produced important results both in the experimental and theoretical fields, making him arguably the greatest all-around physicist of the twentieth century.

Let us return now to the second half of the nineteenth century. At the time of Planck's youth physics, was an important science, but there was no reason to regard it as the most fundamental of all the sciences. If any discipline was thought to have this position, it was chemistry. How did it come about that physics arrived at the position of being the most basic science? We shall discuss this question in the next chapter.

3

Science

Chemistry as the Fundamental Science

Chemistry as a science, distinct from the magic of alchemy, had been founded in the seventeenth century by Robert Boyle and developed by the end of the eighteenth to a high point by Antoine Lavoisier, whose life was cut short on the guillotine during the French Revolution. Its stage at the dawn of the nineteenth century was such that Humphry Davy had good reasons to regard it as the most fundamental of all the sciences. In his popular lectures at the Royal Institution Davy certainly argued strongly that it was. Here are some excerpts from his lecture given on January 21, 1802 (Humphry Davy, Collected Works, vol. 2, page 311):

> Chemistry is that part of natural philosophy which relates to those intimate actions of bodies upon each other, by which their appearances are altered, and their individuality destroyed.
>
> This science has for its objects all the substances found upon the globe. It relates not only to the minute alterations in the external world, which are daily coming under the cognizance of our senses, and which in consequence, are incapable of affecting the imagination, but likewise to the great changes, and convulsions in nature, which, occurring but seldom, excite our curiosity, or awaken our astonishment.

> The phenomena of combustion, of the solution of different substances in water, of the agencies of fire; the production of rain, hail, and snow, and the conversion of dead matter into living matter by vegetable organs, all belong to chemistry ....

And on page 312:

> Natural history and chemistry are attached to each other by very intimate ties. For while the first of these sciences treats of the general external properties of bodies, the last unfolds their internal constitution and ascertains their intimate nature. Natural history examines the beings and substances of the external world, chiefly in their permanent and unchanging forms; whereas chemistry by studying them in the laws of their alterations, develops and explains their active powers and the particular exertions of these powers.

On page 313:

> Even botany and zoology as branches of natural history, though independent of chemistry as to their primary classification, yet are related to it so far as they treat of the constitution and functions of vegetables and animals ....
>
> And in pursuing this view of the subject, medicine and physiology, those sciences which connect the preservation of the health of the human being with the abstruse philosophy of organized nature, will be found to have derived from chemistry most of their practical applications, and many of the analogies which have contributed to give to their scattered facts order and systematic arrangement.

Recall that the young Pasteur chose to pursue chemistry because it was the science generally considered the basis of all other sciences. This was even before new developments transformed chemistry from a mostly qualitative laboratory activity handling Bunsen burners and beakers, mixing gases and liquids, into a quantitative science dealing with numerical measurements. This transformation had been accomplished primarily by the English natural philosopher, John Dalton, by imbuing the ancient notion of atoms with its first scientific basis. If each elementary substance were made up of identical particles — atoms — with a given weight, and if two given elements could be combined into several different compounds — such as different nitrogen oxides — then, he argued, their weight-ratios would have to differ by small whole numbers, as was indeed observed. What is more, it would follow that a chemical compound, consisting of molecules formed from these atoms, would require weight-ratios of its constituent elements that were exactly equal to the weight-ratios of the atoms. (Indeed, the measured weight-ratios required for compounds gave information about the ratios of the weights of the atoms making up a given compound's molecules, as was later verified.) It was the transformation of Democritus's philosophical doctrine of atoms into a scientific hypothesis that made chemists, handling gases, liquids, and solids in their laboratories, and studying their combinations, increasingly dependent upon the use of scales and measurements of weights. In other words, one might say that chemists began acting more and more like physicists.

The second development transforming chemistry was the publication of a table of the elements, ordered by their atomic weights (the atomic weight of an element is the weight of one of its atoms as a multiple of the weight of a hydrogen atom), by the Russian chemist Dmitri Ivanovich Mendeleev in 1869. He had found that if the elements were arranged according to their atomic weights, they showed a certain periodicity of chemical properties, such as their valences. (Valences are the ways in which elements are able to combine with other elements to form compounds.) What is more, elements that are similar in their chemical properties have atomic weights which are nearly the same or else increase in regular steps. All of these regularities he exhibited in what became known as the periodic table of the elements, which listed all of them by increasing atomic weight and simultaneously showed the periodicity of their properties (see Fig. 3.1).

What made his table most convincing was the fact that it contained a number of gaps — places that, according to his rules, should be occupied by elements with properties he listed, but no such substances were known to exist. When these elements were later discovered, with just the predicted atomic weights and chemical features, the great value of Mendeleev's discovery could not be denied. Chemistry had advanced far beyond its stage at the time of Humphry Davy, and his claim that it was the most basic science of them all seemed more justified than ever. The only problem was that, indispensible as Dalton's atoms and Mendeleev's periodic table became for chemists, both notions were entirely *ad hoc* and had neither direct evidence nor reason to justify and understand them.

How Physics Became Most Fundamental

Meanwhile, the science of physics, which had been revolutionized in the seventeenth century by Galileo Galilei and Isaac Newton, was concerned with the motion of objects and matter pushed and pulled by forces, such as the planets attracted to the sun and one another. It also dealt with the motion of fluids, with electricity and magnetism, and with light. These were all subjects that seemed important, but not nearly as basic to other sciences as what Davy had averred for chemistry. What changed all that was the discovery that atoms were not just constructs useful for chemistry but were fundamental even to physics, and that their physical properties would have important implications for all of chemistry. When physicists began to delve even into the interior of atoms, it emerged that the periodic table had a natural physical explanation, and additional results made chemistry completely dependent upon physics. The first of these discoveries occurred during the nineteenth century, though not all at once; the second, also gradually, occurred during the twentieth.

Group → / ↓ Period	1	2	3	4	5	6	7	8	9	10	11	12	13	14	15	16	17	18
1	1 H																	2 He
2	3 Li	4 Be											5 B	6 C	7 N	8 O	9 F	10 Ne
3	11 Na	12 Mg											13 Al	14 Si	15 P	16 S	17 Cl	18 Ar
4	19 K	20 Ca	21 Sc	22 Ti	23 V	24 Cr	25 Mn	26 Fe	27 Co	28 Ni	29 Cu	30 Zn	31 Ga	32 Ge	33 As	34 Se	35 Br	36 Kr
5	37 Rb	38 Sr	39 Y	40 Zr	41 Nb	42 Mo	43 Tc	44 Ru	45 Rh	46 Pd	47 Ag	48 Cd	49 In	50 Sn	51 Sb	52 Te	53 I	54 Xe
6	55 Cs	56 Ba		72 Hf	73 Ta	74 W	75 Re	76 Os	77 Ir	78 Pt	79 Au	80 Hg	81 Tl	82 Pb	83 Bi	84 Po	85 At	86 Rn
7	87 Fr	88 Ra		104 Rf	105 Db	106 Sg	107 Bh	108 Hs	109 Mt	110 Ds	111 Rg	112 Cn	113 Uut	114 Uuq	115 Uup	116 Uuh	117 Uus	118 Uuo

Lanthanides	57 La	58 Ce	59 Pr	60 Nd	61 Pm	62 Sm	63 Eu	64 Gd	65 Tb	66 Dy	67 Ho	68 Er	69 Tm	70 Yb	71 Lu
Actinides	89 Ac	90 Th	91 Pa	92 U	93 Np	94 Pu	95 Am	96 Cm	97 Bk	98 Cf	99 Es	100 Fm	101 Md	102 No	103 Lr

Figure 3.1. A modern version of Mendeleev's periodic table of the elements. The elements Gallium (31) and Germanium (32) were not known at the time but Mendeleev correctly predicted their chemical properties.

The serious exploration of particles as constituents of gases to explain their physical behavior began with attempts to understand the nature of heat. At the beginning of the nineteenth century, the dominant theory of heat was that it consisted of an all-pervading fluid called caloric that intrinsically flowed from warmer bodies to colder ones. On the basis of a variety of observations and experiments demonstrating that heat could be generated by friction — ice could be melted by rubbing, gun barrels became glowingly hot during their manufacture by means of large drill bits, etc. — the caloric theory was gradually replaced by the kinetic theory. It became apparent that heat was nothing but the irregular motion of particles making up gases, liquids, and even solids. For gases, the kinetic theory could be used in its purest form, as no other assumptions were needed — such as forces that held solids and liquids together, giving them more shape and substance than gases — to explain their properties. Nothing but the motion of particles following Newton's laws was required to understand temperature and pressure as well as other features of the behavior of gases.

There was, however, one aspect of the science of heat that was difficult to understand in terms of particles. For the purpose of explaining why heat never flowed from a colder to a warmer body, the notion of entropy had been introduced, and the second law of thermodynamics had been promulgated as a fundamental physical principle: the entropy of a closed system — such as the universe as a whole — could never decrease. But what could entropy possibly mean in terms of the particles making up matter? After all, nothing in a system of particles moving according to Newton's laws of motion was irreversible. How could their motion have a property that might increase in the course of time but could never decrease? The Austrian physicist Ludwig Boltzmann unraveled the mystery.

Boltzmann regarded it as the primary purpose of his life not merely to explain the origin of the second law of thermodynamics in terms of the behavior of atoms, but to convince the world that atoms were real particles and not just abstract models convenient for chemists and physicists. He managed to do the first by means of probabilistic and statistical notions, explicating the meaning of entropy in terms of the probability of a given arrangement of particles of various velocities in a given volume of gas. (For more details, see, for example, my book *What Makes Nature Tick?*) But trying to persuade his fellow scientists that atoms were real and not just figments of theorists' imagination was the fight of his life. They were far too small to be visible under a microscope; nobody had ever seen an atom.

Known nowadays mostly as the eponymous name of the speed of sound, the philosopher-physicist Ernst Mach was Boltzmann's main antagonist. On the basis of his philosophy of positivism, Mach (professor of philosophy at the University of Vienna) taught and argued strenuously that physicists and chemists had no business postulating the existence of such invisible entities as atoms or molecules as parts of reality just because they found them convenient for their theories. It was not until the effects of the irregular motions of water molecules in the form of collisions with tiny dust particles were seen under the microscope that the controversy was finally settled. Called Brownian motion, it was first calculated in detail by Albert Einstein in 1905 and observationally confirmed shortly thereafter. By that time, Boltzmann, who had suffered from depression all his life, had committed suicide.

At this point it could certainly be argued that physics, the science that explained the motion of atoms and molecules, all of which were real particles that chemists could not do without in their theories, was more basic than chemistry. What followed during the next half–century made chemistry even more dependent on physics.

At the beginning of the twentieth century, Ernest Rutherford discovered by ingenious experiments that most of the interior of atoms was empty space. He had told his assistants Hans Geiger and Ernest Marsden to direct a beam of alpha particles (doubly positively charged helium ions) at a thin gold foil, paying special attention to large-angle scattering, and to his great astonishment they found that some of the alpha particles were actually reflected in the backward direction. 'It was almost as incredible as if you fired a fifteen-inch shell at a piece of tissue paper and it came back and hit you,' he said. The only possible explanation was that essentially all the mass of an atom was concentrated in a tiny very dense nucleus in its center, holding a positive electric charge, and a number of the recently discovered negatively charged and extremely light electrons circulated around like the planets around the sun, attracted to the nucleus by the electrostatic force instead of gravity. (Their number had to be such as to exactly neutralize the whole.) This surprising but intuitively appealing picture of what an atom looked like, however, was incompatible with Maxwell's firmly accepted electromagnetic theory, because the circulating electrons would have to constantly radiate light, lose energy, and crash into the center. Such an atom could not be stable. So Rutherford's young Danish assistant Niels Bohr, a man of a distinctly revolutionary temperament, solved the problem by simply postulating that electrons in certain orbits with specified energies could remain there without radiating, whether this violated Maxwell's theory or not. The quantum theory of the atom was thus born in 1912, about seven years after Einstein had initiated the quantum theory of light.

When Bohr's *ad hoc* quantum rules were fleshed out by Werner Heisenberg, Erwin Schrödinger, and Paul Dirac into what became known as quantum mechanics, it became clear that Mendeleev's periodic table followed as a direct consequence of the new physics. As the electric charge of the nuclei of successive atoms in the table increased, so did the corresponding number of electrons in them. More and more of these electrons were forced to circulate in larger and larger orbits, because Wolfgang Pauli had added the exclusion principle, which forbade more than two electrons from occupying a single orbit — we need not concern ourselves here with the question why two rather than one, which has to do with the 'spin' of the electron — and the chemical properties of a given element depended entirely on the outermost electrons of its atoms. (For more details about the quantum theory, see for example my book, *What makes Nature Tick?.*)

Eventually even the interior of the nucleus became subject to physical investigation, and the neutron was discovered, an electrically neutral particle whose weight differed very little from that of the electrically positive proton, the nucleus of hydrogen and the lightest of the elements. Nuclei were found to consist of both. The number of protons in the nucleus of a given atom determined the position of the corresponding element in the periodic table — that number being equal to the number of electrons, as the electric charge of an electron was exactly equal and opposite in sign to that of a proton — whereas the total number of 'nucleons,' i.e., protons and neutrons together, determined its atomic weight as a multiple of the weight of hydrogen, whose nucleus consisted of just one proton. It also became clear that many elements had more than one isotope, that is, forms that had identical chemical properties and were thus indistinguishable by chemical means, but yet had slightly different atomic weights. The nuclei of isotopes simply contained the same number of protons — hence the same number of electrons in the atom and identical chemical properties — but different numbers of neutrons. The form in which many elements are found in nature consists, in fact, of a mixture of several of their isotopes. This explains why the atomic weights of these mixtures are not exactly whole numbers, which had puzzled chemists for a long time.

As time went on, the equations of quantum mechanics became the backbone of all basic calculations of chemists, particularly after the work of the prominent chemist Linus Pauling. His book (written in collaboration with his student E. B. Wilson) *Introduction to Quantum Mechanics with Application to Chemistry*, was one of the earliest textbooks on the new subject, and his later book, *The Nature of the Chemical Bond*, transformed chemistry almost into a sub-discipline of physics. The intricate and multiple ways in which different elements bonded in their enormous variety of compounds were all shown to be the result of the arrangements of electrons in their atoms and the physical forces they exerted on one another. There could be no question but that physics had dethroned chemistry as the most fundamental science.

Indeed, by means of quantum mechanics, the explanatory power of physics reached everywhere. All other properties of solid and fluid matter beyond the reach of chemistry, such as electric and heat conductivity, the newly discovered phenomena of superfluidity and superconductivity at extremely low temperatures, as well as the properties of semiconductors — materials that conduct electricity not as well as conductors like copper or aluminum but that are not insulators either — were explained in terms of the young theory. What is more, just as Newton had extended physics to the realm of the planets by explaining their orbits by means of the universal force of gravity, so Einstein's general theory of relativity vastly increased that realm to include the stars and galaxies. The shining of the Sun as well as the slow cooling of the Earth (recall the importance of this for Darwin's theory of evolution) became understood by means of nuclear physics, and so were even the observed abundances of all the elements in the universe. The entire historical development of the cosmos as a whole, including its expansion in the course of time, were explicated by a combination of general relativity, quantum field theory, and the experimental discovery of a large number of formerly unknown elementary particles. The new discipline of astrophysics, with its reach to the stars, is intimately dependent upon the discoveries made by means of enormous particle-accelerators in Europe and the United States. (See, for example, Lisa Randall's recent book *Knocking on Heaven's Door.*)

There can now be no doubt that of all the sciences, physics is not only the one with the greatest reach, but also that in some basic sense all other sciences, directly or indirectly *via* chemistry, depend on it to various degrees. This does not mean, of course, that biologists have to learn quantum field theory or keep up with the latest discoveries of more particles. The new discipline of biophysics makes use of only certain very restricted aspects of physics. But Humphry Davy was correct in his insistence that biology is basically dependent on chemistry. Now that chemistry has become dependent upon physics, biology has inherited that dependency as well.

On Reductionism

The argument I have presented above began with Humphry Davy arguing that chemistry was the fundamental science in the sense that biologists required it for understanding their science. This was followed by showing that, in view of the progress in physical discoveries concerning atoms, one could not fully comprehend chemistry without physics. In other words, in some sense biology was reduced to chemistry and chemistry to physics. Some commentators attach the label 'reductionism' to this line of argument, and attack it vociferously. All of science should form a harmonious whole, they argue, each discipline of inquiry resting on its own way of looking at the world rather than having to rely on the results and conclusions of other disciplines. This approach is called holistic, which originates from the Greek word for "whole".

I disagree. To reduce a newly discovered phenomenon to others already well understood or less complicated is the essence of what we mean by an explanation. We do not explain a strange happening by introducing a new, unfamiliar entity or let it stand on its own. However, it may be a good idea to clarify what such a reduction means, and especially what it does not mean.

There have been famous instances in the history of science when it was thought to be necessary to introduce new and different entities to do justice to certain complex phenomena. Perhaps the best known of these was the very old notion of vitalism in biology, whose origins go back to ancient Egyptian philosophy. It postulated that in the grand scheme of nature anything having to do with life was fundamentally distinguished from inorganic matter by the presence of an intrinsic vital force. This idea survived well into the nineteenth century. When one of its invalidators, the chemist Friedrich Wöhler, used nothing but inorganic components to synthesize urea, a waste product of living organisms, he described it as 'the slaying of a beautiful hypothesis by an ugly fact.' Some of today's opponents of reductionism regard any elimination of the need for proliferation of new entities specific to limited areas of science as a regrettable diminution of the special character of that area: if chemistry is reduced to physics, it loses its own intrinsic way of thinking, which should be part of the holistic discipline of science. Such a fear, however, is quite unjustified.

Consider the example of the notion of entropy introduced into thermodynamics in order to deal with the phenomenon of irreversibility: heat flows from a hot to a cold body but never the other way around. The second law of thermodynamics decreed that, for a closed system, the entropy, which had been defined in purely thermodynamic terms, could increase but never decrease. Boltzmann managed to reduce the strange idea of entropy to the behavior of any physical system consisting of an extremely large number of particles, each of which behaved according to completely reversible laws, and yet as a whole turned out to behave irreversibly.

This kind of phenomenon, in which the behavior of a system consisting of well understood entities exhibits quite new and unexpected features under certain specific circumstances — in this case, that the number of particles is extremely large — has come to be called an *emergent property*. Such emergent properties have been found to be of particular importance in understanding complexity and the many new features of complex physical systems. The strange newly discovered phenomenon of superconductivity was explained by means of Cooper pairs, emergent entities explained in terms of the conduction of electrons moving at very low temperature in the lattice of molecules constituting a metal. The important notion of emergent properties serves as an explanation only if the unexpected properties are causally understood on the basis of what is already known. If they are merely postulated *ad hoc* they are no better than assuming a miracle.

The existence of emergent properties shows that the reduction of one discipline of phenomena to another does not necessarily rob it of its own distinct character. Boltzmann did not destroy thermodynamics, and specialists in thermodynamics still use their own, very effective way of understanding and dealing with the phenomena in their field. What is more, thermodynamics remains a well-integrated part of physics. Similarly, chemists are not deprived of their special chemical tools just because they ultimately have to rely on physics for a basic understanding and detailed calculations. Nor do biologists relinquish any of their life-related insights by the fact that chemistry and physics lie at the bottom of their science. Pasteur was not handicapped by relying on his knowledge of chemistry for many of his medical insights. Nor did Linus Pauling lose his admirable chemical intuition by understanding the basic nature of the chemical bond in terms of quantum mechanics. (Indeed, when Francis Crick and James Watson raced to be first to explain the chemical structure of the DNA molecule, which lies at the very foundation of life, their closest competitor was the quintessential chemist Linus Pauling. For another good example see the recent book edited by Sloan and Fogel, *Creating a Physical Biology*.)

Of course, reductionism can be misused by the members of one scientific discipline for the purpose of regarding themselves as intellectually superior to those of another. (See Chapter III, entitled 'Two Cheers for Reductionism' in Steven Weinberg's well-known book *Dreams of a Final Theory* for a discussion of this phenomenon.) However, the notion that reductionism diminishes the freedom of specially adapted ways of intuition of individual disciplines is misconceived. Obscuring what it seeks to illuminate, holism in science is a philosophical doctrine with no explanatory value.

We have come, then, to the final conclusion that the most basic science, the discipline that has both the widest reach and the deepest foundation, is physics. It reaches from the understanding of the fundamental particles existing in the world to the evolution of the universe, and it provides the basis on which chemistry and biology rest. Its practitioners, the physicists, have reason to view with awe the grandeur of nature and to feel humility about contributing to its explanation.

Science and mathematics are among the highest achievements of the human intellect. Centuries from now, let us hope that it will not be the ruins of the Pentagon in Washington that we leave behind us as our equivalent of the Parthenon in Athens. I expect that it will be our mathematics and science, with physics as its basis, rather than war memorials that will command admiration long after we are gone.

References and Further Reading

Cliff, Nigel, *Holy War: How Vasco da Gama's Epic Voyages Turned the Tide in a Centuries-Old Clash of Civilizations.* New York: Harper Collins Publ. Co., 2011

Darwin, Charles, *The Annotated Origin: A Facsimile of the First Edition of the Origin of Species.* (Annotated by James T. Costa). Belknap Press of Harvard University Press, 2009

Debré, Patrice, *Louis Pasteur.* The Johns Hopkins University Press, 1998

Fermi, Laura, *Atoms in the Family: Life with Enrico Fermi, Architect of the Atomic Age.* University of Chicago Press, 1954

Grant, Edward, *A History of Natural Philosophy: From the Ancient World to the Nineteenth Century.* Cambridge University Press, 2007

Heilbron, J. L., *The Dilemmas of an Upright Man: Max Planck as Spokesman for German Science.* University of California Press, 1986

Holmes, Richard, *The Age of Wonder: How the Romantic Generation Discovered the Beauty and Terror of Science.* New York: Pantheon Books, 2008

Iltis, Hugo, *Life of Mendel.* (translated by Eden and Cedar Paul), New York: Hafner Publishing Co., 1966

Knight, David, *Humphry Davy: Science and Power.* Oxford: Blackwell Publishers, 1994

Levathes, Louise, *When China Ruled the Seas: The Treasure Fleet of the Dragon Throne 1405–1433,* New York: Simon & Schuster, 1994

McCalman, Ian, *Darwin's Armada: Four Voyages and the Battle for the Theory of Evolution.* New York: W. W. Norton & Co., 2009

Newman, William R., *Atoms and Alchemy: Chymistry & the Experimental Origins of the Scientific Revolution.* University of Chicago Press, 2006

Newton, Roger G., *What Makes Nature Tick?* Harvard University Press, 1993

Newton, Roger G., *The Truth of Science: Physical Theories and Reality.* Harvard University Press, 1997

Randall, Lisa, *Knocking on Heaven's Door: How Physics and Scientific Thinking Illuminate the Universe and the Modern World.* New York: Harper Collins Publ. Co., 2011

Sloan, P. F. and B. Fogol, editors, *Creating a Physical Biology: The Three Man Paper and Early Molecular Biology.* University of Chicago Press, 2011

Sootin, Harry, *Gregor Mendel: Father of the Science of Genetics.* New York: The Vanguard Press, In., 1959

Terrall, Mary, *The Man Who Flattened the Earth: Maupertuis and the Sciences of the Enlightenment.* Chicago: University of Chicago Press, 2002

Thomas, John Meurig, *Michael Faraday and the Royal Institution: The Genius of Man and Place.* Bristol: Adam Hilger, 1991

Tyndall, John, *Faraday as a Discoverer.* New York: Thomas Y. Crowell Co., 1961

Weinberg, Steven, *Dreams of a Final Theory.* New York: Pantheon Books, 1992

Williams, L. Pearce, *Michael Faraday: A Biography,* New York: Da Capo Press, Inc., 1965

Illustration Credits

Figure 1.1:   From L. Levathes, *When China Ruled the Seas*, p. 21.

Figure 1.2:   US Naval Historical Center Photograph, from geology.com.

Figure 1.3:   From Holmes, *The Age of Wonder*, page facing page 141.

Figure 2.1:   From Wikipedia.org.

Figure 2.2:   From Iain McCalman, *Darwin's Armada*.

Figure 2.3:   From history.nih.gov/exhibits/nirenberg/popup_htm/01_mendel.htm

Figure 2.4:   From Hugo Iltis, *Life of Mendel*, plate XII.

Figure 2.5:   Photograph by Pierre Lamy Petit. Wikimedia Commons.

Figure 2.6:   Image Albumen silver print on card taken by Patrice Debr Petit, Harvard Art Museum.

Figure 2.7:   Picture at the Royal Institution; from L. Pierce Williams, *Micheal Faraday*, plate 14a.

Figure 2.8:   From J. M. Thomas, *Michael Faraday and the Royal Institution*, p. 32.

Figure 2.9:   From J. M. Thomas, *Michael Faraday and the Royal Institution*, p. 47.

Figure 2.10:   From Scientistsinformation.blogspot.com.

Figure 2.11:   From Los Alamos Photo Laboratory/Science Photo Library. Portrait of Enrico Fermi (1901–1954) the Italian–American physicist who contributed much to the field of quantum statistical mechanics.

Index

abundances, 80
action at a distance, 63
aerobes, 52
aesthetics, 33, 34
alchemy, 26, 71
Alexander, 8
algebra, 26
Ampère, André Marie, 61
Amundsen, Roald, 17, 21
anaerobes, 52
anaerobic germs, 55
anthrax, 55
antiseptics, 55
Apollo 11, 24
Archimedes, 25
Aristotle, 8, 27
Armstrong, Neil, 24
astrology, 6, 7
astrophysics, 80
atomic weight, 73, 79
atoms, 25, 73, 76
attenuated virus, 57
Aztecs, 16

Babylonians, 4, 6
bacteria, 52
Banks, Joseph, 16
Barton, Otis, 22
bathyscaphe, 22
bathysphere, 22
Beagle, 36
beauty, 32, 34
Beebe, Charles William, 22
benzine, 61
Beowulf, 4

Bhagavad Gita, 4
Big Dipper, 6
biochemistry, 52
biophysics, 81
black body, 65
Bohr, Niels, 29, 66, 78
Boltzmann, Ludwig, 29, 65, 76–82
Bondi, H., 33
Boyle, Robert, 26, 27, 29, 71
Bridgman, Percy, 35
Bruno, Giordano, 3
Buddha, 2

calendar, 5
Captain Cook, 16
chemistry, 26
China, 6
Christian Europe, 2
Christianity, 4
chromosomes, 48
civilization, 1
classical physics, 65
Columbus, Christopher, 13
compass, 11
complexity, 82
conductivity, 80
Confucianism, 2, 3
constellations, 6
Cook, Frederick, 21
cooling of the Earth, 80
Cooper pairs, 82
Copernicus, 3
Correns, Carl Erich, 50
Cortés, Hernán, 16
cosmology, 80

Crick, Francis, 43, 83
Curie, Marie, 29, 36

da Gama, Vasco, 14, 15
Dalton, John, 73
Darwin, Charles, 29, 36, 43, 49
Davy, Humphry, 59, 71–72
de Vries, Hugo, 50
de Maupertuis, Pierre Louis Moreau, 18
Democritus, 25
diamagnetic, 64
Dirac, Paul, 29, 78
dog star, 6
dominant traits, 46
dynamos, 63

eclipses, 6
Eddington, Arthur, 19
Edison, Thomas, 29
Einstein, Albert, 19, 29, 31–33, 66
electric motors, 63
electrolysis, 61
electromagnetic field, 63
electromagnetic induction, 62
electromagnetic rotation, 61
electromagnets, 63
elegance, 34
emergent properties, 82, 83
entropy, 76, 77, 82
ephemeris, 6, 15
Ericson, Leif, 9
exclusion principle, 78
explanation, 28, 81

facts, 30
Faraday, Michael, 29, 33, 36, 59–65
Faustus, 26
fermentation, 52
Fermi, Enrico, 29
ferromagnetic, 64
FitzRoy, Robert, 36

Galapagos Islands, 38
Galileo, 3, 27, 29, 74
gangrene, 54

Gauss, 29
Gautama, 2
Geiger, Hans, 77
Gemini, 6
general theory of relativity, 19, 80
germs, 52
Gibbs, 29
Gilgamesh, 3
giraffes, 12
gnomon, 4
Greek civilization, 6

Hardy, G. H., 29, 34
heat, 76
Heisenberg, Werner, 29, 66, 78
Helmholtz, 29
Henry, Joseph, 62
Herchel, Caroline, 23
Herschel, William, 23
Hippocrates, 30
holism, 84
holistic, 81, 82
Homer, 3
Hooker, Joseph, 42
Huxley, Thomas, 35, 42
hybridization, 44

iguanas, 38
Iliad, 3
Incas, 16
Indian civilization, 2
induction, 31
Institut Pasteur, 58
intelligent design, 42
irreversibility, 82
Islam, 3, 4
isotope, 79

Jainism, 2
Jenner, Edward, 56
Judaism, 4

Kelvin, 29, 42
Kepler, 3
Keynes, John Maynard, 31

kinetic theory, 76
Koch, Robert, 55

Lagrange, 29
Lamarckian explanations, 49
Laplace, 29
Lavoisier, Antoine, 29, 71
Leonardo da Vinci, 27
lines of force, 63
Lister, Joseph, 54
Livingstone, David, 20
Lorentz, 29
Lupus, 6
Lyell Charles, 39

Mach, Ernst, 77
Magellan, Ferdinand, 15
magnetochemistry, 64
Malthus, Thomas, 39
Manifesto of 93, 66
Marsden, Ernest, 77
Maxwell, 29
Maxwell, James Clerk, 63
Medawar, Peter, 32
Meister, Joseph, 57
Mendel's Law of Segregation, 48
Mendel's second law, 49
Mendel, Gregor, 28, 36, 42–49
Mendeleev, Dmitri Ivanovich, 73
microbes, 52
Ming emperors, 10
modern science, 26
Modern Synthesis, 43
molecules, left-handed, 52
molecules, right-handed, 52
Muller, Joseph, 29

natural selection, 39, 40
Needham, Joseph, 2
neutron, 79
Newton, Isaac, 27–29, 31, 74
Nibelungen, 4
Noether, Emmy, 32
North Star, 6
Northwest Passage, 17

nucleon, 79

Odyssey, 3
Ørsted, Hans Christian, 61

paramagnetic, 64
Park, Mungo, 19
particle-accelerators, 80
Pasteur, Louis, 29, 30, 36, 51–58, 83
pasteurization, 53
Pauli, Wolfgang, 78
Pauling, Linus, 79, 83
Peary, Robert, 21
periodic table, 73, 75, 78
phases of the Moon, 5
photons, 66
Piccard, August, 22
Piccard, Jacques, 22
Pinta, 14
Pizarro, Francisco, 16
Planck's constant, 66
Planck's Law, 66
Planck, Max, 29, 36, 65–67
Plato, 8
platypus, 38
Poincaré, 29
polarization of light, 63
polarized light, 51
Polo, Maffeo, 9
Polo, Marco, 9, 10
Polo, Niccolò, 9
prediction, 32
proton, 79
Ptolemy, 25
pure science, 25, 26
Pythagoras, 7, 25

qilin, 12
quanta, 66
quantum mechanics, 78
quantum theory, 66
quantum theory of radiation, 66

rabies, 56
radioactivity, 42

Ramanujan, Srinivasa, 34
recessive traits, 46
reductionism, 81, 82
reliability, 34
Ross, James Clark, 18
Ross, John, 17
Royal Institution, 59, 63
Rutherford, Ernest, 29, 33, 77

Saint Elmo's fire, 12
Santa Clara, 14
Santa Maria, 14
Schrödinger, Erwin, 29, 78
science wars, 35
Secchi, Pietro, 24
second law of thermodynamics, 76, 82
semiconductors, 80
Semmelweis, Ignaz, 54
sex cells, 47
Shackleton, Ernest, 21
silkworms, 54
Sirius, 6
smallpox, 55
special theory of relativity, 66
spectacles, 12
spontaneous generation, 53
spores, 55
Stanley, Morton, 20, 21
stereo-chemistry, 52
Sun, 80
sun clocks, 5
superconductivity, 80, 82
superfluidity, 80
syphilis, 14

Taoism, 2, 3
Thales, 7
thermodynamics, 65
transformers, 63
treasure ships, 10
Trieste, 22
Trojan war, 3
Tschermak, Erich, 51

Ursa Major, 6

vaccine, 55, 56
valences, 73
validity, 34
van Leeuwenhoek, Antonj, 52
Vikings, 4
virus, 56
vitalism, 81

Wöhler, Friedrich, 82
Wallace, Alfred Russel, 41
Walsh, Donald, 22
water clocks, 5
Watson, James, 43, 83
Weinberg, Steven, 83
Wien's Law, 65, 66
Wien, Wilhelm, 65
Wilson, E. B., 79

Zheng He, 11–13
Zhu Di, 11
Zhu Gaozhi, 12
Zhu Qizhen, 13
Zhu Zhanji, 13